Communications
in Computer and Information Science 1972

Series Editors

Gang Li⊙, *School of Information Technology, Deakin University, Burwood, VIC, Australia*
Joaquim Filipe⊙, *Polytechnic Institute of Setúbal, Setúbal, Portugal*
Ashish Ghosh⊙, *Indian Statistical Institute, Kolkata, West Bengal, India*
Zhiwei Xu, *Chinese Academy of Sciences, Beijing, China*

Rationale

The CCIS series is devoted to the publication of proceedings of computer science conferences. Its aim is to efficiently disseminate original research results in informatics in printed and electronic form. While the focus is on publication of peer-reviewed full papers presenting mature work, inclusion of reviewed short papers reporting on work in progress is welcome, too. Besides globally relevant meetings with internationally representative program committees guaranteeing a strict peer-reviewing and paper selection process, conferences run by societies or of high regional or national relevance are also considered for publication.

Topics

The topical scope of CCIS spans the entire spectrum of informatics ranging from foundational topics in the theory of computing to information and communications science and technology and a broad variety of interdisciplinary application fields.

Information for Volume Editors and Authors

Publication in CCIS is free of charge. No royalties are paid, however, we offer registered conference participants temporary free access to the online version of the conference proceedings on SpringerLink (http://link.springer.com) by means of an http referrer from the conference website and/or a number of complimentary printed copies, as specified in the official acceptance email of the event.

CCIS proceedings can be published in time for distribution at conferences or as post-proceedings, and delivered in the form of printed books and/or electronically as USBs and/or e-content licenses for accessing proceedings at SpringerLink. Furthermore, CCIS proceedings are included in the CCIS electronic book series hosted in the SpringerLink digital library at http://link.springer.com/bookseries/7899. Conferences publishing in CCIS are allowed to use Online Conference Service (OCS) for managing the whole proceedings lifecycle (from submission and reviewing to preparing for publication) free of charge.

Publication process

The language of publication is exclusively English. Authors publishing in CCIS have to sign the Springer CCIS copyright transfer form, however, they are free to use their material published in CCIS for substantially changed, more elaborate subsequent publications elsewhere. For the preparation of the camera-ready papers/files, authors have to strictly adhere to the Springer CCIS Authors' Instructions and are strongly encouraged to use the CCIS LaTeX style files or templates.

Abstracting/Indexing

CCIS is abstracted/indexed in DBLP, Google Scholar, EI-Compendex, Mathematical Reviews, SCImago, Scopus. CCIS volumes are also submitted for the inclusion in ISI Proceedings.

How to start

To start the evaluation of your proposal for inclusion in the CCIS series, please send an e-mail to ccis@springer.com.

Valarmathi K. · Ramathilagam A. ·
Sankarganesh Seeni · Utku Kose ·
Bui Thanh Hung · Kottilingam Kottursamy
Editors

Artificial Intelligence and Digital Twin Technology

1st International Conference, IconAIDTT 2023
Sivakasi, India, April 26–28, 2023
Proceedings

Editors
Valarmathi K.
PSR Engineering College
Virudhunagar, Tamil Nadu, India

Sankarganesh Seeni
PSR Engineering College
Virudhunagar, Tamil Nadu, India

Bui Thanh Hung
Industrial University of Ho Chi Minh City
Ho Chi Minh City, Vietnam

Ramathilagam A.
PSR Engineering College
Virudhunagar, Tamil Nadu, India

Utku Kose
Süleyman Demirel University
Isparta, Türkiye

Kottilingam Kottursamy
Anna University
Chennai, Tamil Nadu, India

ISSN 1865-0929　　　　　　　ISSN 1865-0937　(electronic)
Communications in Computer and Information Science
ISBN 978-3-031-77798-1　　　ISBN 978-3-031-77799-8　(eBook)
https://doi.org/10.1007/978-3-031-77799-8

© The Editor(s) (if applicable) and The Author(s), under exclusive license
to Springer Nature Switzerland AG 2024

This work is subject to copyright. All rights are solely and exclusively licensed by the Publisher, whether the whole or part of the material is concerned, specifically the rights of translation, reprinting, reuse of illustrations, recitation, broadcasting, reproduction on microfilms or in any other physical way, and transmission or information storage and retrieval, electronic adaptation, computer software, or by similar or dissimilar methodology now known or hereafter developed.
The use of general descriptive names, registered names, trademarks, service marks, etc. in this publication does not imply, even in the absence of a specific statement, that such names are exempt from the relevant protective laws and regulations and therefore free for general use.
The publisher, the authors and the editors are safe to assume that the advice and information in this book are believed to be true and accurate at the date of publication. Neither the publisher nor the authors or the editors give a warranty, expressed or implied, with respect to the material contained herein or for any errors or omissions that may have been made. The publisher remains neutral with regard to jurisdictional claims in published maps and institutional affiliations.

This Springer imprint is published by the registered company Springer Nature Switzerland AG
The registered company address is: Gewerbestrasse 11, 6330 Cham, Switzerland

If disposing of this product, please recycle the paper.

Preface

We are delighted to present the revised list of 13 accepted and presented papers for the IconAIDTT 2023 (International Conference on Artificial Intelligence & Digital Twin Technology-2023) Conference Proceedings, now published in the Communications in Computer and Information Science (CCIS) series by Springer. This collection of papers showcases cutting-edge research and technological advancements in various fields of information and communication technology. Each contribution has undergone rigorous peer review to ensure the highest standards of quality and innovation.

IconAIDTT 2023 was held at PSR Engineering College, Tamil Nadu, India, on April 26–28, 2024. Overall 41 papers were presented at the conference, and from these, 12 regular and 1 short paper were selected for publication in this volume. Every paper was selected based on the comments received from three different reviewers, and additional revision was carried out before the final camera-ready submission. The topics covered reflect the diversity and depth of current research, including security and performance enhancements in wireless sensor networks, detection of position falsification attacks in VANETs, and AI-based Covid-19 detection using modified ResNet models. Other notable works include the application of blockchain for data security, the development of interactive road safety systems using smart helmets, a bibliometric analysis of customer relationship management systems, and wireless safety systems for underground mine workers. Additionally, the proceedings explore energy-based routing protocols in underwater sensor networks, IoT-based smart agriculture monitoring, age-based content display using face recognition, detection of student engagement levels using ResNet, heart disease prediction using deep belief networks, and handwriting analysis for bank cheque verification using EfficientNet.

We extend our gratitude to the authors for their valuable contributions and to the reviewers for their diligent work in evaluating the submissions. We would also like to express our sincere thanks to the keynote speakers, whose insights and expertise greatly enriched the conference:

- Srinath Doss, Dean, Faculty of Engineering and Technology, Botho University, Botswana
- Saravanakumar Gurusamy, Professor, Federal TVET Institute, Addis Ababa, Ethiopia

We are confident that this collection of papers will serve as a significant resource for researchers, practitioners, and students, inspiring further innovation and development in the field. The breadth of topics and the innovative approaches presented in these papers highlight the dynamic and evolving nature of information and communication technology. We hope that the insights and findings shared in these proceedings will be both informative and stimulating, fostering new ideas and advancements in the years to come.

Sincerely,

September 2024

Valarmathi K.
Ramathilagam A.
Sankarganesh Seeni
Utku Kose
Bui Thanh Hung
Kottilingam Kottursamy

Organization

Program Chairs

K. Valarmathi	PSR Engineering College, India
Utku Kose	Suleyman Demirel University, Turkey
Bui Thanh Hung	Thu Dau Mot University, Vietnam
K. Kottilingam	Anna University, India
K. Valarmathi	PSR Engineering College, India
A. Ramathilagam	PSR Engineering College, India

International Technical Advisory Board

Atta ur Rehman Khan	Ajman University, UAE
Ali Kashif Bashir	Manchester Metropolitan University, UK
Jey Veerasamy	University of Texas, Dallas, USA
Satheesh	Sunchon National University, South Korea
Keping Yu	Waseda University, Japan
Mehdi Shadaram	University of Texas at San Antonio, USA
Zing-Ming Zhong	Tsinghua University, China
Korhan Cengiz	Trakya University, Turkey
Xavier Fernando	Ryerson University, Canada
P. D. D. Dominic	Universiti Teknologi Petronas, Malaysia
Sheeba Backia Mary Baskaran	Huawei Technologies, Sweden
A. S. M. Kayes	La Trobe University, Australia

National Technical Advisory Board

Muthurajkumar	Anna University, MIT, India
Nandhakumar R.	VIT, India
Janakiraman	Pondicherry University, India
Kishore Kumar	NIT Warangal, India
Murugan K.	VIT, India
Preethi	Anna University Coimbatore, India
Purusothaman	GCT Coimbatore, India
Sudhakar T.	VIT, India
Vinayak Shukla	CMC Vellore, India
Yamin Khan	Microsoft, India

Organizing Committee

M. P. Pushpalatha	PSR Engineering College, India
M. Ilayaraja	PSR Engineering College, India
P. Srihari	PSR Engineering College, India
R. Gokulraj	PSR Engineering College, India
M. Thamaraiselvi	PSR Engineering College, India
M. Vijay	PSR Engineering College, India
S. Gopalakrishnan	PSR Engineering College, India
N. Jayanthi	PSR Engineering College, India
A.T. Kalidoss	PSR Engineering College, India
A. Vivek	PSR Engineering College, India

Reviewers

J. Roscia Jeya Shiney	PSR Engineering College, India
N. Janakiraman	K.L.N. College of Engineering, India
N. Arivazhagan	SRMIST, Chennai, India
C. BagathBasha	Kommuri Pratap Reddy Institute of Technology, India
G. Karthikeyan	PSR Engineering College, India
R. Palanikumar	PSR Engineering College, India
S. Gopalakrishnan	Vel Tech University, India
Hariharan	SRMIST, India
K. Vimala Devi	Vellore Institute of Technology, India
P. Ranjithkumar	PSR Engineering College, India
S. Priyadarsini	PSR Engineering College, India
S. Ramesh	JAIN (Deemed-to-be University), India
Singaravelan S.	PSR Engineering College, India
S. Rajasoundaran	SRMIST, India
R. Aruna	PSR Engineering College, India
S. Rajaprakash	AVIT, VMRF, India
K. Vimala Devi	SRMIST, India
Jayanthi	SRMIST, India
K. Ramalakshmi	PSR Engineering College, India
R. Muniraj	PSR Engineering College, India
J. Sivadasan	PSR Engineering College, India
R. Ramani ASP	PSR Engineering College, India
B. Manjurathi	PSR Engineering College, India
S. Krishnaveni	PSR Engineering College, India
M. Gengaraj	PSR Engineering College, India

L. Krishna Kumari	PSR Engineering College, India
M. Vimala	PSR Engineering College, India
P. Lingeswari	PSR Engineering College, India
Sudhakar T.	Anna University, India
Senthilkumar	SRMIST, India
Raja	VIT, India

Additional Reviewers

Anandh	Chitra
Kumar	Rajkumar
Senthil	Rajasekar
Sudhakar	Ponnusamy
Murugan	Suresh
Nandhakumar	Dinesh
Valaramathi	Kesavan
Praveen	Thiraviyam

Contents

Advanced AI and IoT Solutions for Environmental, Healthcare, and Security Challenges

A Novel Approach for Enhancing the Security and Performance of Wireless Sensor Networks With a Single Cluster with Multi-hop Communication 3
    S. Gopalakrishnan, D. Hemanand, A. Gnana Soundari, S. V. Hemanth, Mohammad Aljanabi, and J. Jasmine Hephzipah

A Hybrid Machine Learning Model for Position Falsification Attacks for Intrusion Detection in VANET 14
    G. Jeyaram, V. Vidhya, M. Madheswaran, and R. Shirley Jeeva Malar

Covid-19 Detection Using AI Deep Modified Resnet Model from Human Chest X-ray Images 24
    Narenthira Kumar Appavu and Nelson Kennedy Babu

A Blockchain with RB-BM23-1 Method Used to Secure the Analyzed Data 34
    Kishore, S. Rajaprakash, K. Karthik, Neha, Sarangakrishna, and Pavan Chandra

Development of an Interactive Road Safety System Using a Smart Helmet for Bike Users to Avoid Bike Accidents 43
    S. Rajalingam, S. Kanagamalliga, K. Sakthi Priya, R. Karpaga Priya, and S. Kavitha

Customer Relationship Management a Decision Support System: Bibliometric Analysis 1990–2023 55
    S. Lokesh and S. Vasantha

Deep Learning Techniques for Wireless Networks and Data Analysis

Wireless Sensor Network-Based Wireless Safety System Using Underground Mine Workers 71
    S. Gopalakrishnan, K. Rani Swetha, S. Manisha, and P. Sreenath Reddy

A Meticulous Analysis on Energy Based Routing Protocol in Underwater Sensor Networks 82
    B. Anandha Mathavan, A. Shenbagharaman, B. Paramasivan, and B. Shunmugapriya

An IOT Based Smart Agriculture Monitoring System for Precision Farming ... 103
   K. Vivekrabinson, A. Vishnu, S. Jeya Aravinth,
   and P. Sundara Mahalingam

AGE Based Content Display by Using Face Recognition 112
   Ahmad Dayoub, Ali Hasan, Sangya Bhandari, Poojitha Ghandhupu,
   and K. S. Arvind

IoT Enabled Heart Disease Accuracy Prediction of Healthcare Dataset
Using Deep Belief Network 131
   Rahama Salman and Subodhini Gupta

Handwriting Analysis for Bank Cheque Verification Using EfficientNet 138
   Jaydeep Ranpariya, Roshan Saravanan, Alvin James, Pranav Abraham,
   and S. Ramesh

An Intensive Approach to Solve Linguistic Issue Using Data Mining
and Ontology Based Advanced Algorithms 148
   M. M. Uma Maheswari and N. Arivazhagan

Author Index 163

Advanced AI and IoT Solutions for Environmental, Healthcare, and Security Challenges

A Novel Approach for Enhancing the Security and Performance of Wireless Sensor Networks With a Single Cluster with Multi-hop Communication

S. Gopalakrishnan[1]([✉]), D. Hemanand[2], A. Gnana Soundari[3], S. V. Hemanth[4], Mohammad Aljanabi[5], and J. Jasmine Hephzipah[6]

[1] Department of Electronics and Communication Engineering, Veltech Rangarajan Dr.Sagunthala R&D Institute of Science and Technology, Chennai 600062, Tamil Nadu, India
drsgk85@gmail.com
[2] Department of Computer Science and Engineering, S.A. Engineering College, Thiruverkadu 600077, Tamil Nadu, India
[3] Department of Computer Science and Engineering, Saveetha School of Engineering, SIMATS, Thandalam, Chennai 602105, India
[4] Department of Computer Science and Engineering, Hyderabad Institute of Technology and Management, Medchal-Malkajgiri 501401, Telangana, India
[5] Department of Computer, College of Education, Alirqia University, Baghdad, Iraq
[6] Department of Electronics and Communication Engineering, R.M.K Engineering College, RSM Nagar, Kavaraipettai 601206, Tamil Nadu, India
jjh.ece@rmkec.ac.in

Abstract. It is common practice to employ wireless sensor networks (WSNs) for intercontinental conversations. Quality of service (QoS), which may be quantified in terms of end-to-end latency, energy efficiency, and packet delivery ratio, is crucial for these wireless networks. Using a routing protocol with distinguishing features is advised to enhance service quality. Due to the absence of an alternate way in the event of a path failure, the single-path routing mechanism in WSN has significant difficulties in providing a continuing communication channel. This issue is resolved by implementing a multipath routing method using a clever protocol. WSNs are dispersed throughout space and are autonomous. WSNs are susceptible to security concerns since they are deployed sporadically and lack a central authority. Using the attacker identification Protocol, a more effective security methodology has been created based on the approach that has been provided (AIP). In this study, we also suggest the SCMC protocol, which increases bandwidth, packet delivery ratio, and end-to-end delay in order to enhance QoS. The single cluster head Multi-hop Communication Protocol stabilizes the energy level to achieve excellent energy conservation and a longer lifespan. For multi-hop communications, a single cluster may produce several dynamic multi-paths among its members. Testing of the proposed work against the ECMP and SPEED protocols is done in Ns-2. According to the findings, the suggested protocol performs better than the other two protocols regarding shortest latency, minimal energy consumption, and improved delivery ratio.

Keywords: WSN · Security · Attacker Identification Protocol · Single Cluster · Multi-hop Communication

1 Introduction

A Wireless Sensor Network (WSN) is a decentralized network that connects various geographical locations using effective technologies. According to certain routing algorithms, sensor nodes are dispersed randomly throughout the network and communicate with one another over wireless networks. The main reason for creating these WSN is to track and analyze various environmental and physical characteristics. They are pollutants, pressure changes, temperature variations, and noise pollution, all of which have a substantial impact on a nation's development. In addition to monitoring natural events, these WSNs are also used for smart spaces, noise detection, inventory traffic, terrorist movements, seismic detection, and medical monitoring. A communication path was constructed inside a network covering a wide area by connecting thousands of nodes through intermediary nodes. It is difficult to build a huge network on a worldwide scale because it requires overcoming a number of obstacles, including limited power, high failure rates, and bandwidth. Due to the complexity of achieving network leadership and the consequent need to design robust and energy-efficient protocols, the system may prioritize doing so. Each sensor node in a network should make the most of its available power source. Data accuracy may vary widely depending on the kind of sensor used and the scenario at hand. Problems may be avoided by using these records indefinitely with little power consumption. There has to be coordination between nuclear reactors and data collecting to maximize efficiency and extend the life of the network. Therefore, it

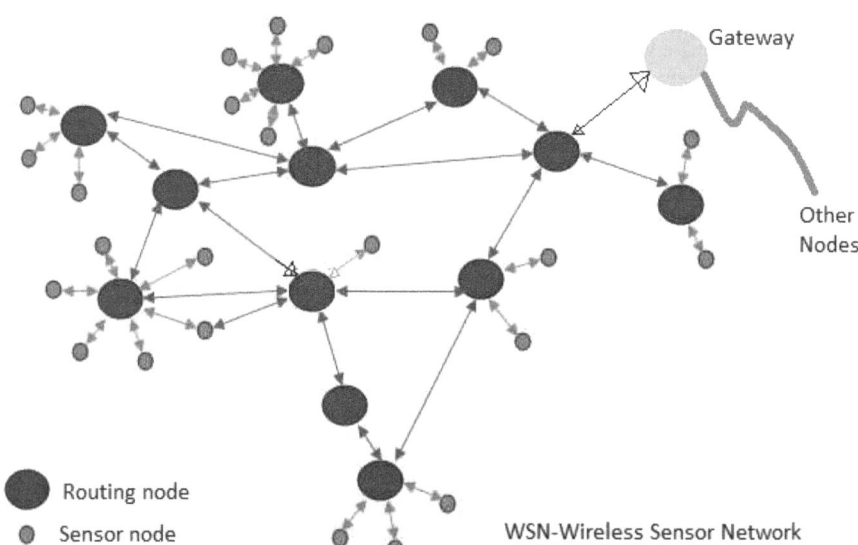

Fig. 1. Wireless Sensor Network Model

is of interest that a protocol be devised that accounts for a wide variety of important factors, including application and network density. The majority of existing WSN routing algorithms were originally created with the special features of WSNs in mind (Fig. 1).

In this research, we use multipath-based, QoS-based, cluster-based, and query-based methods to design our protocol. Here is the strategy we recommend: For data transmission, the Single Cluster with Multi-hop Communication (SCMC) protocol uses several paths. To further improve the system's speed and network lifetime, a cluster approach might be used. The following is the outline for this document: Sect. 2 discusses the literature review and other related methods. In Sect. 3, we explore the significance of multipath routing strategies and the recommended methodology for cluster-based and event-based WSNs. In Sect. 4, we detail the mechanics of our multipath routing technique. The fifth part provides a visual comparison and evaluation of the suggested solution to the current gold standard. Section 6 provides a synopsis of the paper.

2 Literature Survey

Quos-based routing systems are covered in this section [1]. We extensively addressed several of the publications that are pertinent to our proposed approach rather than summarizing the associated work. The initial routing system that supported QoS in those early days was the Sequential Assignment Routing (SAR) protocol [2]. Using fuzzy logic control (FLC) based prediction algorithms, Gautam et al. (2018) presented a Security-Aware Dual Authentication based Routing (SDAR) to strengthen network security and protect nodes from threats.

A cluster-based QoS aware routing system [3] by K. Akkaya and M. Younis handles actual time and non-real-time traffic using a queuing model. When establishing links, it only takes into account endpoint delay using a cost function. Using end-to-end restrictions, kleast-cost path analysis determines the optimal path between two points. The most efficient route will be used to send data. All nodes share bandwidth. Because of this, expanding the size of the network's capacity is unnecessary. This protocol's primary problem is that it fails to account for transmission latency and cost calculation, resulting in network end-to-end delays.

Speed [4] At the time, a well-known QoS-based routing system promised to address real-time end-to-end latency issues. The optimal paths are calculated using location data relayed by sensor nodes. To be efficient, packets have to be sent within the time period specified, and end-to-end latency must be determined by spreading the total distance depending on the nodes' packet delivery rates. The problem is that when the network is overloaded, it will result in network jamming.Sensor nodes may be deployed in dense arrays for a range of applications due to their low cost and compact size [9, 10]. As a result of diverse applications, wireless sensor network traffic may comprise time-sensitive signals and reliability-demanding packets.

To accomplish QoS in WSNs, Felemban et al. [5] propose the Multi-path and Multi-Speed Routing Protocol. (MMSPEED). It achieves varied QoS in the timeliness domain by using several channels that transmit data dependent on the dependability of the domains. In their paper [6], Huang and Y. Fang introduced the concept of multi constrained QoS multi-path routing (MCMP) protocols. When transmitting data, these protocols choose the most efficient routes feasible. Here, service quality is evaluated based on factors like reliability and delay time. An optimization problem based on linear integer programming is used to tackle end-to-end latency issues. To meet QoS requirements with minimal power consumption, it employs a technique that conforms to the ECMP protocol's strict limits on the number of hops and the amount of energy used. The second protocol is a multipath routing protocol named EQSR. It is both energy-efficient and QoS-aware, and it assures that all nodes in a network have the same energy usage [12]. Homogeneous clustering methods overlook the possibility of node heterogeneity because they presume that all sensor nodes have about the same amount of energy. By modifying this strategy, we may propose a heterogeneous clustered system for energy-efficient wireless sensor networks. Whichever node has the highest weighted election probability based on residual energy is elected as the cluster leader [11].

The energy that is consumed for the transmission of single bit of data is equal to the energy consumed to do 1000 operation in the SN (Anastasi et al. in 2015). Accordingly, there is larger energy consumption by the communication sub-system. Consequently, the protocols for the forwarding and energy efficient routing must be deployed for communication. In WSN, there might be redundant dataset collection by the neighbourhood nodes from the environment. For making the reduction of the processing energy and the data forwarding volume to the sink node, an approach for clustering to be used. Few clusters make up the majority of the SN in the network. Embarrassingly communicated data are passed from the SN to the sink node. Every cluster has a cluster head, whose primary duty is energy conservation. The cluster's duty is occasionally shifted to the other SN of the cluster in an effort to lengthen the cluster lifespan. This clustering approach is used by the LEACH protocol, which stands for the Low Energy Adaptive Clustering Hierarchy (Heinzelman et al. in 2014). Thousands of SN operations may be performed with the same amount of energy needed to transport a single bit of data (Anastasi et al. in 2015). Accordingly, there is larger energy consumption by the communication sub-system. Consequently, the protocols for the forwarding and energy efficient routing must be deployed for communication. In WSN, there might be redundant dataset collection by the neighbourhood nodes from the environment. For making the reduction of the processing energy and the data forwarding volume to the sink node, an approach for clustering to be used. There are a limited amount of clusters made up of all the SN in the network. It is annoyingly sent the data that has to be sent from the SN to the sink node. Every cluster has a cluster head whose primary duty is energy conservation. Regularly rotating the cluster's responsibility to the other SN extends its lifespan. LEACH (Low Energy Adaptive Clustering Hierarchy) uses this clustering method. (Heinzelman et al. in 2014).

When SN are mobile nodes, LEACH is upgraded to LEACH Mobile (Kim and Chung in 2016). Whenever there is a movement of the SN between the cluster or leave one cluster and joins another cluster. The cluster membership is redefined, the schedule of time for reaching the newer location is confirmed in Time-division multiple access schedule of the moving node. All decision systems, require accurate and complete domain information. In case of the information being uncertain, Fuzzy Logic (FL) is used for making decisions (Godbole in 2017). FL uses the rule sets to make judgments in real time. Finding the optimal connection between two nodes in a network is a common use of the evolutionary algorithm. (Nagib and Wahied in 2018). The goal function seeks the path between origin and destination that uses the least amount of energy. An efficient energy management system is developed using a FL genetic technique. The overall amount of hops to the destination, energy at each node, and packets to be transported influence fuzzy rules. Genetic algorithms selected this groundbreaking standard.

Dutta et al. (2018) provided a FL controlled cluster head selection algorithm for WSN. Initially, for selecting the cluster head the fuzzy controller is designed. The cluster head must have the energy efficiency and needed nodes in neighbourhood. So, the node centrality, neighbourhood density, and residual energy are the input parameters. The likelihood of that node being chosen as the cluster leader is the outcome variable. Density and residual energy in the surrounding area may be described using the linguistic terms high, medium, and low. The linguistic variables for thenode centrality is taken as far, adequate and far. According to the information on WSN and the input variables If, then rules were derived. This rule of fuzzy is used during the selection of cluster head.

An improved version of the LEACH protocol called as Leach Mobile protocol is derived when SN are the mobile nodes (Kim and Chung in 2016). Whenever there is a movement of the SN between the cluster or leave one cluster and joins another cluster. The cluster membership is redefined, the schedule of time for reaching the newer location is confirmed in the Time-division multiple access schedule of the moving node. Accurate and comprehensive domain information is essential for all decision systems.Fuzzy Logic (FL) is utilized for decision making when there is ambiguity in the available data. (Godbole in 2017). FL uses the rule sets to make judgments in real time. Finding the optimal connection between two nodes in a network is a common use of the evolutionary algorithm. (Nagib and Wahied in 2018). The goal function seeks the path between origin and destination that uses the least amount of energy. To have energy management effectively, a FL genetic approach is derived. A node's energy reserves, the distance to the destination, and the number of hops in between each affect the kinds of fuzzy rules that may be applied. This cutting-edge rule was chosen utilizing a genetic technique.

The explanation of a well-known clustering method may be found below [7]. The cluster head is constructed from randomly selected sensor nodes created using this method. Following standard operating practice, data was sent straight from the central station to the cluster leaders. The long distance between cluster heads and base stations causes delays even when LEACH addresses the energy gap. HEED [8] is a popular clustering-based routing method that chooses the cluster head based on residual and reference energy. To obtain high data speeds, a wireless Mobile Ad-Hoc network (MANET) must increase packet access efficiency [13–16]. The detection of harmful nodes will make it difficult to lessen the severity because of the similarities between trustworthy and malicious nodes in the sensing region.

2.1 Significance of the Research

WSNs are wireless networks that are individually designed and require minimal infrastructure for communication and status monitoring. By carefully choosing the cluster head, the suggested research project may offer the optimal transmission system and platform for actuation. By effectively utilizing energy optimization techniques, the lifetime of the WSN is also more stable than the previous technologies. The suggested elliptic curve cryptographic system also satisfies the criteria of high security with least restricted resources. As a result, a cyber-secured solution must strictly adhere to the node's authentication, data confidentiality, resilience, and non-compromise against cyberattacks.

3 Proposed System Architecture

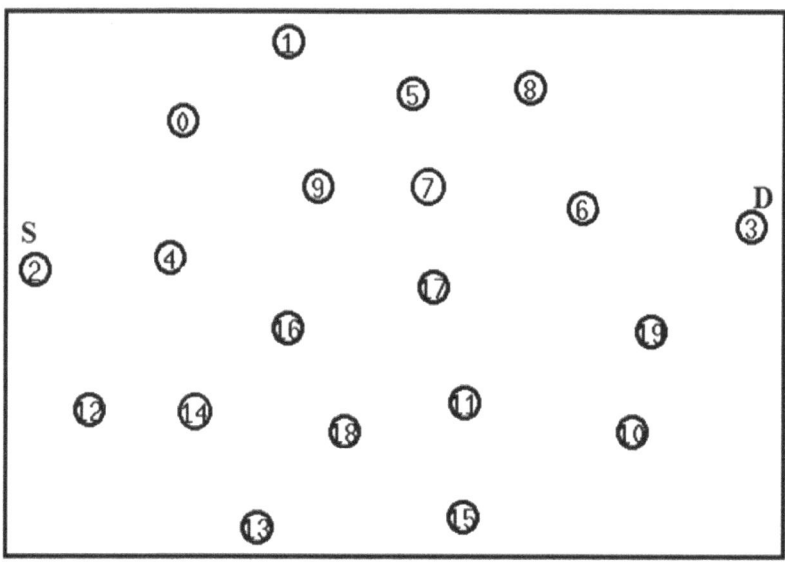

Fig. 2. Sensor Node Deployment Dynamic Structure

Numerous low-cost, flexible sensor nodes form a sensor network. These may be located in close proximity to one another utilizing a network of nodes equipped with detection, packet transmission, and data processing capabilities, or they can be dispersed over a vast area using sensor media. Applications [1] imply that their size is more suitable for use. Information in WSNs is sent between nodes wirelessly. Data collection typically involves either a sink node or a gateway node. It has a dispersed architecture and has several sensor nodes [2] (Fig. 2).

The nearby nodes of each sensor node exchange information with one another. A perfect node deployment plan may eliminate issues with routing, data fusion, and communication, among others. Node deployment is a significant problem in WSN. The smallest quantity of energy can be used to increase a WSN's longevity. We found that placing nodes uniformly decreases complexity, enhances manageability, and produces a greater range of homogeneity. The majority of WSN apps assigned these nodes at random. Because wrong node placement increases the complexity of WSN. Energy consumption is a major problem for WSNs that may be solved to increase their lifespan [3].

4 Proposed Design Goal

This work creates a novel SCMC protocol for achieving QoS across several clusters. The enhancement of bandwidth, packet delivery ratio, and round-trip delay are its key objectives. Our method results in a shared cluster head across all randomly connected nodes. Each node's energy level will affect the overall performance of this single dynamic cluster head. The leader of the cluster will serve as the source node, sending information to the other nodes in proportion to their available power. Everyone inside the cluster head's range receives the same consistent amount of energy. Once a cluster head is chosen, all nodes send data to it. Following the suggestions will save energy and extend longevity. The most useful element is reassessing each cluster node's energy level relative

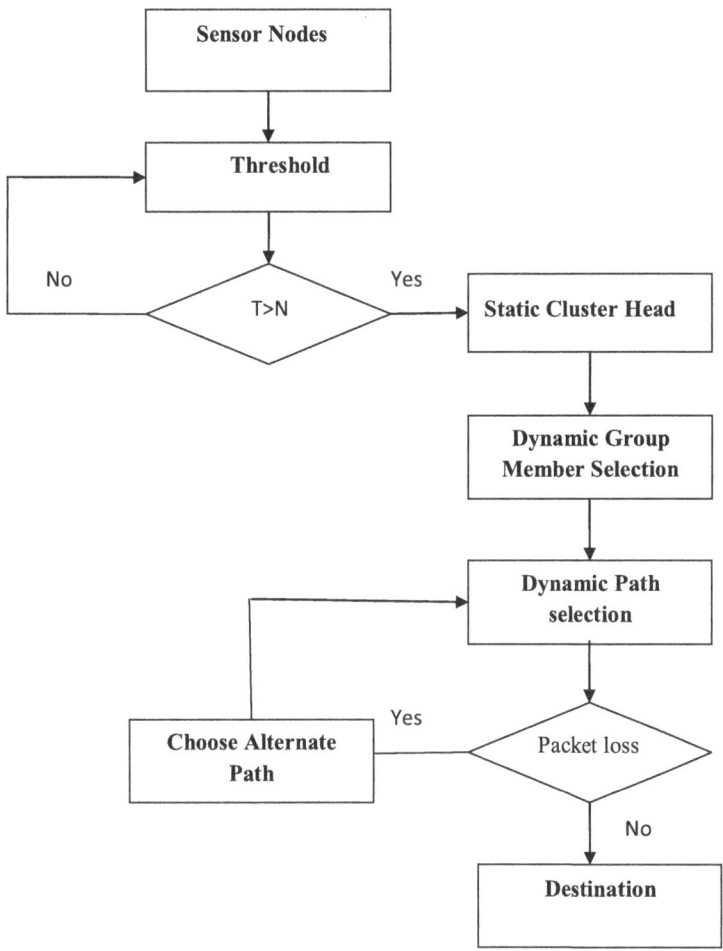

Fig. 3. Flow chart for Proposed Scheme

to the others. In order to improve transmission performance and efficiency, higher-energy nodes were employed for communication and energy savings. More details on the proposed procedure are provided below.

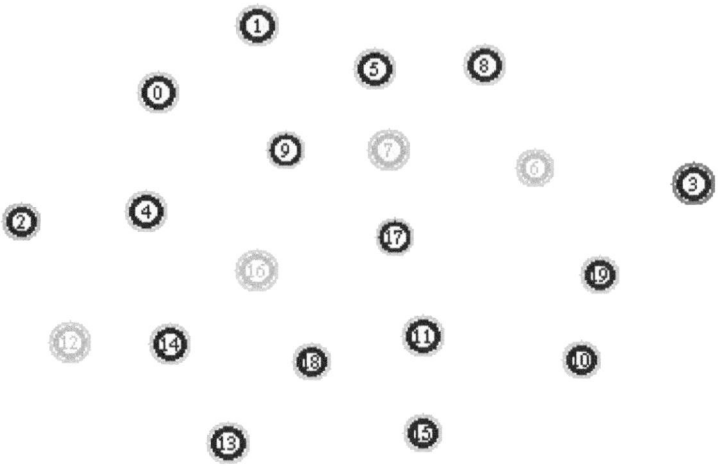

Fig. 4. Dynamic multi cluster 12, 16, 7, 6

As can be seen in Fig. 3, a cluster is created by connecting the nodes 2, 12, 16, and 3 with the help of the proposed Multipath Cluster with Dynamic Mode Routing Protocol. The node 2 serves as the cluster's source node, while the node 3 serves as the cluster's destination node. It is possible for information to be sent in the sequence presented from one node in the cluster to the next (Fig. 5).

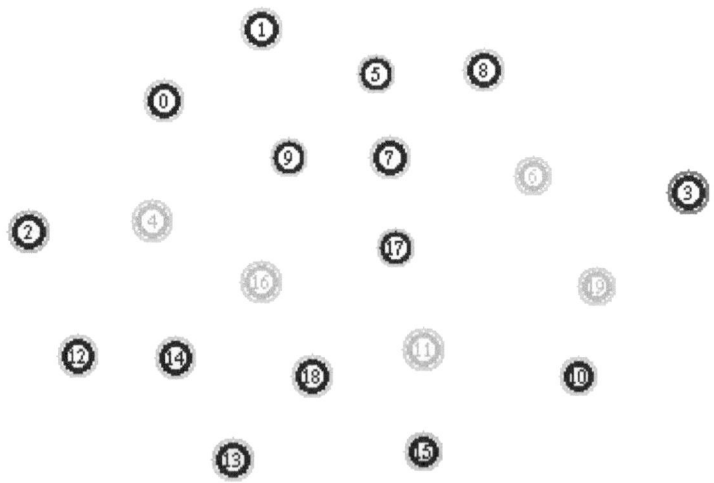

Fig. 5. Dynamic multi cluster 4, 16, 11, 6, 19

In the event of a cluster failure, the set of nodes 4, 16, 11, 6, and 19 that share the same source node 2 and destination node 3 will automatically form a dynamic cluster, as seen in Fig. 4. Figure 4 depicts the dynamic formation of a new cluster after the expiration of an earlier cluster.

5 Experimental Results

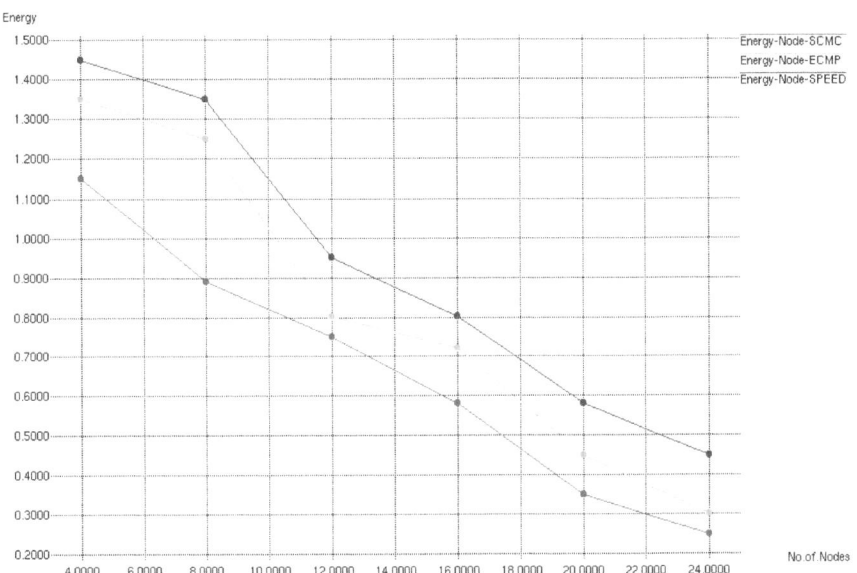

Fig. 6. Number of Nodes Vs Energy

Figure 6 shows packet transmission energy vs node count. 24 nodes relay packets using various protocols. SCMC is the most power-efficient protocol.

Figure 7 compares MCDMRP, ECMP, and SPEED end-to-end latency as a function of packet delivery. Dynamic clustering is the fastest. The other two methods increase delay linearly with data transmission rate. The graph shows that our technique beats ECMP and SPEED.

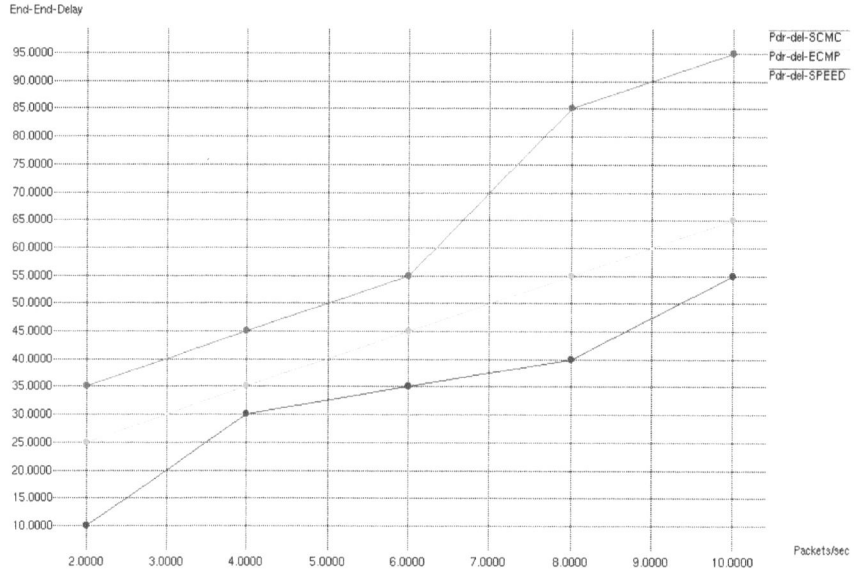

Fig. 7. Packets/sec Vs End-to-end delay

6 Conclusion

Many writers in this study spoke about how to implement real-time QoS in WSN. Finally, by using the source link as the cluster head and generating various paths dynamically, we develop a novel protocol that provides a multi-hop communication protocol. Even if a dynamic path was constructed, communication would stop if a path or connection failed. In this path configuration, the cluster leader will function as the source node, sending information to the other nodes according to their available resources. The suggested method will dynamically build multipath in the case of a path failure. The efficiency of the suggested system is shown by contrasting it with ECMP and SPEED in terms of end-to-end latency, throughput, and packet delivery ratio. Graphical depictions of the results show that our SCMC is exceptional and much exceeds the ECMP and SPEED. Therefore, the highest Quality-of-Service (QoS) outcomes may be achieved using our suggested strategy for wireless sensors.

Conflict of Statement. All authors there is no Conflict of Intrest to publish this Article.

References

1. Karthick, P.T., Palanisamy, C.: Optimized cluster head selection using krill herd algorithm for wireless sensor network. Automatika **60**(3), 340–348 (2019). https://doi.org/10.1080/00051144.2019.1637174
2. Middya, R., Chakravarty, N., Naskar, M.K.: Compressive sensing in wireless sensor networks – a survey. IETE Tech. Rev. **34**(6), 642–654 (2017). https://doi.org/10.1080/02564602.2016.1233835

3. Jiang, N., Zhou, R., Ding, Q.: Dynamics of wireless sensor networks. Int. J. Distrib. Sens. Netw. **5**(6), 693–707 (2009). https://doi.org/10.1080/15501320802581565
4. Chaaran, K.N., Younus, M., Javed, M.Y., BasedMultisink, N.S.N.: Minimum delay energy efficient routing in wireless sensor networks. Eur. J. Sci. Res. **41**(3), 399–411 (2010)
5. Sinde, R., Begum, F., Njau, K., Kaijage, S., Shih, K.-P. (eds.): Lifetime improved WSN using enhanced-LEACH and angle sector-based energy-aware TDMA scheduling. Cogent Eng. **7**(1) (2020). https://doi.org/10.1080/23311916.2020.1795049
6. Haque, Md.E., Matsumoto, N., Yoshida, N.: Context-aware cluster-based hierarchical protocol for wireless sensor networks. Int. J. Ad Hoc Ubiquitous Comput. **4**(6), 379–386 (2009). https://doi.org/10.1504/IJAHUC.2009.028666
7. Saad, E.M., Awadalla, M.H., Darwish, R.R.: Adaptive energy-aware gathering strategy for wireless sensor networks. Int. J. Distrib. Sens. Netw. **5**(6), 834–849 (2009). https://doi.org/10.1080/15501320903235400
8. Huang, X., Fang, Y.: Multiconstrained QoS mutlipath routing in wireless sensor networks. Wirel. Netw. **14**(4), 465–478 (2008). https://doi.org/10.1007/s11276-006-0731-9
9. Krishnan, M., Lim, Y.: Reinforcement learning-based dynamic routing using mobile sink for data collection in WSNs and IoT applications. J. Netw. Comput. Appl. **194** (2021). https://doi.org/10.1016/j.jnca.2021.103223. PubMed: 103223
10. Kumar, D., Aseri, T.C., Patel, R.B.: EEHC: energy efficient heterogeneous clustered scheme for wireless sensor networks. Comput. Commun. **32**(4), 662–667 (2009). https://doi.org/10.1016/j.comcom.2008.11.025
11. Sohrabi, K., Gao, J., Ailawadhi, V., Pottie, G.J.: Protocols for self-organization of a wirless sensor network. IEEE Pers. Commun. **7**(5), 16–27 (2000). https://doi.org/10.1109/98.878532
12. Shahzad, M.K., Cho, T.H.: An energy-aware routing and filtering node (ERF) selection in CCEF to extend network lifetime in WSN. IETE J. Res. **63**(3), 368–380 (2017). https://doi.org/10.1080/03772063.2016.1241721
13. Younis, O., Fahmy, S.: HEED: a hybrid, energy-efficient, distributed clustering approach for ad hoc sensor networks. IEEE Trans. Mob. Comput. **3**(4), 366–379 (2004). https://doi.org/10.1109/TMC.2004.41
14. Al-Rousan, M., Kullab, D.: Real-time communications for wireless sensor networks: a two-tiered architecture. Int. J. Distrib. Sens. Netw. **5**(6), 806–823 (2009). https://doi.org/10.1080/15501320903048647
15. Borkar, G.M., Mahajan, A.R.: Security aware dual authentication based routing scheme using fuzzy logic with secure data dissemination for mobile ad-hoc networks. J. Appl. Secur. Res. **13**(2), 223–249 (2018). https://doi.org/10.1080/19361610.2017.1387737
16. Subburayalu, G., Duraivelu, H., Raveendran, A.P., Arunachalam, R., Kongara, D., Thangavel, C.: Cluster based malicious node detection system for mobile ad-hoc network using ANFIS classifier. J. Appl. Secur. Res. (2021). https://doi.org/10.1080/19361610.2021.2002118

A Hybrid Machine Learning Model for Position Falsification Attacks for Intrusion Detection in VANET

G. Jeyaram[1(✉)], V. Vidhya[1], M. Madheswaran[2], and R. Shirley Jeeva Malar[1]

[1] M.E.T Engineering College, Nagercoil, Tamilnadu, India
jeyaramgj@gmail.com
[2] Muthayammal Engineering College, Rasipuram, Tamilnadu, India

Abstract. The majority of self-driving vehicles are susceptible to various types of attacks because of their dynamic network along with communication design topology. These kinds of vehicles are dependent on outside VANET communication bases. Industry and academia have shown a lot of interest in it, but road safety, traffic congestion and security have not been correctly addressed in fresh centuries. Building a safe framework for the VANET communication scheme along with being able to sense various kinds of attacks are the greatest significant requirements of network security, have been adequately studied by numerous students, to address these issues. This work suggests a Hybrid DSVM scheme that is based on Support Vector Machine (SVM) and Decision Tree (DT) Machine Learning algorithms to build a secure framework to detect attack, to improve performance as well as adapt to the VANET scenario. The exploratory outcomes show that this approach gives the improved outcomes when contrasted with various AI based Calculations to identify assault.

Keywords: vehicular ad-hoc networks · machine learning · Decision Tree · Support Vector Machine · attacks

1 Introduction

VANET is one of the greatest significant insightful transport advancements that has cleared the way to various requests essentially connected with wellbeing of streets and our day-to-day exercises, for example, cautioning mishaps, limiting public property harms, and giving data about gridlock, climate, and Web office [1]. The following are the five main categories into which attacks in VANETs can be divided: Attacks on Availability, Accountability, and Authentication, as well as Attacks on Integrity and Confidentiality [2]. Mobile ad hoc networks are used in a unique way in vehicular ad hoc networks, or VANETs. They use monitored roads and autonomous vehicles for safer intelligent transportation [3].

To resolve these problems, it is expected to fabricate safe structure for the correspondence framework in VANET along with to recognize various sorts of assault some the main necessities of the organization safety, which has been concentrated on satisfactorily by numerous specialists [4]. A Secret Markov Model channel model is given reasonable for IDS under VANET to diminish the heap as well as location time short of negotiating the precision of recognition. To rapidly screen messages from neighbouring vehicles (NVs), the filter model predicts their normal or abnormal future behaviour [5].

A hybrid optimization-rooted Deep Maxout Network (DMN) for attack classification in VANET is developed. The designed hybrid optimization algorithm is utilized to choose and route the Cluster Head (CH) [6]. Labelled intrusion datasets as well as Machine Learning (ML) techniques that some readily available to the public are briefly discussed. The implementation of NIDS in a variety of networking scenarios, including traditional networks, wireless sensor networks (WSNs), cloud networks and Internet of Things (IoT) networks, is later explained [7].

In VANETs, false distributed denial-of-service (DDoS) attacks along with data injection attacks, particularly stealthy DDoS attacks, are taken into consideration [8]. It is presented a ML mechanism that enhances position falsification attacks for the performance of IDS by taking benefit of three new features primarily linked the contributor position. In addition, it compares two distinct ML methods for classification i.e., Random Forest (RF) as well as k-Nearest Neighbour (kNN), both of which some utilized to identify malicious vehicles by making use of these features [9]. In order to create a distributed Intrusion Detection System (IDS) that makes use of ML techniques, this work related to Cooperative Intelligent Transport Systems (C-ITS) makes a feature combination and then makes use of those combination. Using a large, custom-built dataset, the proposed misbehaviour detection method is compared to other methods to determine its performance. The following are the work's contributions:

- Some features are added to make it easier to spot position falsification errors. The evaluated distance among the sender as well as the receiver, the difference among the declared, the evaluated angle of arrival and the evaluated distance among sender as well as receiver are all position-related features. The detection capabilities are improved by these features.
- The proposed features are exploited by a distributed IDS mechanism. Whenever it receives a new communication, all vehicle performs the trained model after central training has been completed.
- DSVM is an ensemble learning technique that combines two ML techniques to boost the IDS's performance by combining the marks of various base learners.
- To validate the success of the proposed work, a comprehensive analysis of various attack types and traffic densities is carried out.

The despite of this work are as follows: The related recent works are in Sect. 2, Sect. 3, Workflow and methodology; Sect. 4, Discussion and Results of the Experiment; Sect. 5, Conclusion.

2 Literature Review

The most recent research on IDS in VANET is described in detail in this section. ML, DL, or a combination of both ML and DL approaches has all been used in the past to develop a number of methods. Table 1 arranges the current methodologies which are evaluated regarding their strategies, geniuses, and their cons.

Table 1. Literature review

Paper and author	Performance Measures		
	Method	Advantages	Disadvantages
Bangui et al. [10]	Hybrid ML model	Decrease ML problems	Does not work in real-time
Karthiga et al. [11]	Effective Intelligent Intrusion Detection System (IDS)	Secure and reliable communication	No optimized performance
Alsarhan et al. [12]	Intrusion detection scheme	Detection a common attack	The system performance is not clear
Arya et al. [13]	Distributed federated learning (FL) based on the intrusion detection	Quality of life, security, and safety	–
Ghaleb et al. [14]	Distributed ensemble learning based on the concept of C-ITS	Give about their traffic situation, movement state, along with road conditions	IDS model will be examined with supervised as well as unsupervised ML techniques
Marwah et al. [15]	A unique ML technique	Unique ML technique to improve VANET's effectiveness	Does not optimize the parameters

Hybrid ML models, FL, and ensemble methods were previously proposed for IDS in VANET. The problems with machine learning can be solved using these techniques, as can the detection of attacks and secure communication. However, these approaches lack the ability to optimize the parameters and have some drawbacks when working with real-time. Therefore, a hybrid DSVM model with a few features is described in this work to address these issues.

3 Proposed Methodology

Here some vehicle-related features of the proposed IDS mechanism are derived to improve vehicle misbehaviour detection. After that, these features are used to create distributed ML-based IDS that defend against location fabrication attacks in VANETs (Fig. 1).

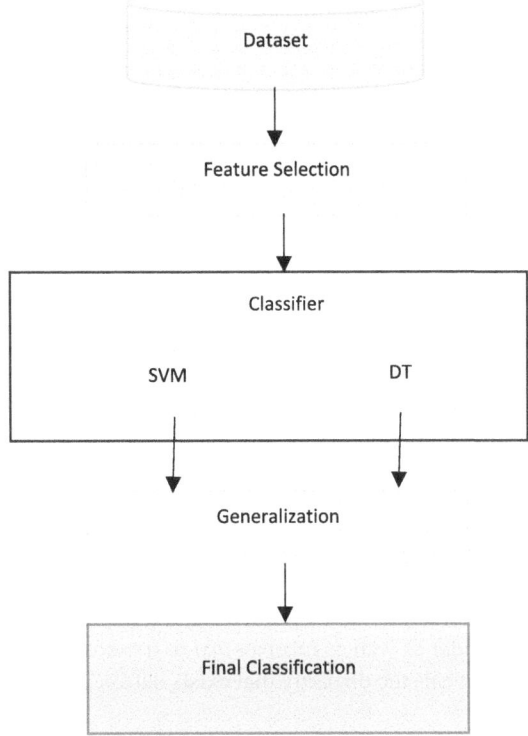

Fig. 1. Proposed IDS Flow

After the centralization of the working out stage, the vehicles are dispersed throughout the detection phase. This work uses SVN and DT, which are typically used to detect misbehaviour and produce satisfying results, to detect malicious vehicles. In order to achieve better classification marks, these two ML techniques will be used together. As a result, the proposed feature combinations are suggested to be implemented using an Ensemble Learning approach. The subsequent subsections will also provide an explanation of the specifics of the learned approaches.

3.1 Feature Selection

The angle of arrival (AoA) is the first new feature. It is obtained by combining the sender-receiver distance with the arctangent function, as shown in Eq. (1).

$$Angle = arctan\frac{\Delta P_{R \to S,y}}{\Delta P_{R \to S,x}} \quad (1)$$

where $P_{R \to S}$ denotes the sender's position relative to the receiver. The evaluated distance between the sender as well as receiver is the second feature that can be used to identify a position falsification attack. Equation (2) shows how the received signal strength indicator (RSSI) can be used to obtain this feature:

$$RSSI = P_{tr} - P_{lo}(dist) \tag{2}$$

where $P_{lo}(dist)$ is the path loss in decibels at the evaluated distance *dist*, and P_{tr} is the transmission power. The transmission power is assumed to be identical across along with constant all nodes. Because of this, no vehicle could alter it. The log-normal shadowing model described in Eq. (3) is then used to estimate the distance.

$$P_{lo}(dist) = P_{lo}(dist_0) + 10 n \log(dist/dist_0) \tag{3}$$

where n is the environment-dependent path loss exponent as well as $P_{lo}(dist_0)$ is the reference power value at a reference distance $dist_0$. The reference power value is measured in dB. In the end, the estimated distance is given by Eq. (4) by utilizing Eqs. (2) and (3):

$$dist = 10^{\frac{P_{tr} - RSSI - P_{lo}(dist_0)}{10n}} * dist_0 \tag{4}$$

The notation $\widehat{dist}$ will be used in the remainder of the paper instead of *dist* because the aim is to value the RSSI-based distance among the sender as well as the receiver. Finally, this study proposes using the difference between the estimated $\widehat{dist}$ along with declared distances between the sender as well as receiver *dist* as a useful feature for misbehavior detection. Equation (5) reveals the disparity that exists between the estimated distances, $|R \rightarrow S|$ and $|R \rightarrow F|$:

$$\Delta dist_{R \rightarrow S} = \left| \sqrt{\Delta P^2_{R \rightarrow S,x} + \Delta P^2_{R \rightarrow S,y}} - \widehat{dist} \right| \tag{5}$$

3.2 Proposed Detection Technique

In C-ITS, the effectiveness of anomaly-based IDS with supervised learning techniques has been demonstrated. Artificial Neural Network (ANN), SVM, Bayesian Classifier, DT along with others is examples of these techniques. Another alternative classification method is ensemble learning, which combines bagging, boosting, and stacking ML classification techniques to enhance the training process. Among all pre-owned ML methods, two of them are for the most part liked and by and large show improved brings about bad conduct discovery: DT and SVM. To avoid the negative things of the organization method that was used, these two methods are suitable for comparing the proposed features to those of others. One of the most well-known ML methods for solving classification issues is SVM. SVM maps the kernel functions using the non-linear information to high-dimensional.

In contrast, DTs are trained using distinct sub-datasets. It classifies the information based on the extreme amount of elections and prevents issues with over-fitting. With additional excluding almost one third of the exercise dataset in all split, these sub-datasets are selected at random. Additionally, the features used in the sub-dataset training are chosen at random.

In order to improve the classification's performance, loading is an ensemble learning technique that enables the integration of diverse knowledge methods. A first organization stage on the equal dataset is performed using a variety of learning strategies in this approach. The predictions of the primary methods are then combined using a meta-classifier or simplification model. The stacking method is used to combine the results in a generalized boosted model (GBM). In order to take advantage of the Stacking method's potential, this paper proposes using GBM for the generalization phase and SVM and DT for the initial step. Using the aforementioned algorithm, we are attempting to identify undesirable behaviour in a network that enables safe communication between autonomous vehicles.

4 Results and Discussions

The dataset is created for the purpose of testing misbehavior detection known as VeReMi which was grounded on a public dataset [16–18] demonstrate the efficiency of the IDS mechanism that has been proposed and to make it easier to compare it to previous works. Obtaining a simulated dataset or real as well as rearranging it by same senders as well as receivers is the first step in this work. Then, a few features are determined to be utilized for the vehicles grouping as aggressors or typical vehicles by utilizing diverse transfer learning strategies along with ML. Lastly, the marks that were obtained some related to those that were produced using features that have already been proposed by other investigators. The VeReMi dataset that has been made public contains five distinct attacks and takes into account three distinct attacker rates (10%, 20%, as well as 30%) for various traffic densities like medium, low, along with high.

The exhibition of the proposition is assessed utilizing exactness, accuracy, review, and F1-Score. The confusion matrix, comprises of the metrics like the true positive (TP), true negative (TN), false positive (FP) as well as false negative (FN), must first be obtained before these performance indicators can be calculated. The overall right organization relation, also known as the amount of correct calculation of both 0 and 1

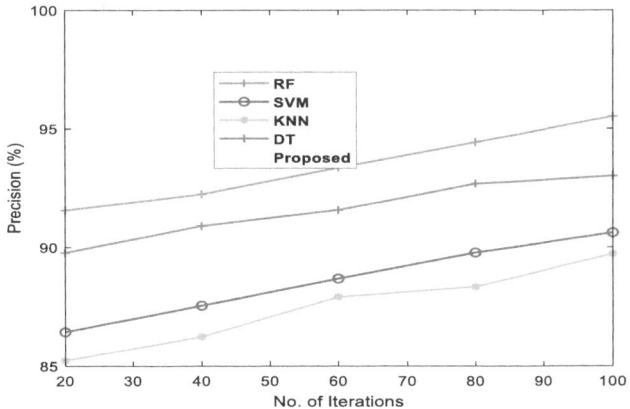

Fig. 2. Precision Comparison Graph

for each case, is provided by accuracy. Recall is the actual amount of TP to every case, while precision is the amount of TP to every case predicted to be 1. Recall as well as precision has a negative correlation: Recall will decrease as precision increases, and the other way around. F1-Score is a trade-off indicator that must be used to balance recall and precision. To resolve this issue, F1-Score analyses the harmonic normal of recall along with precision.

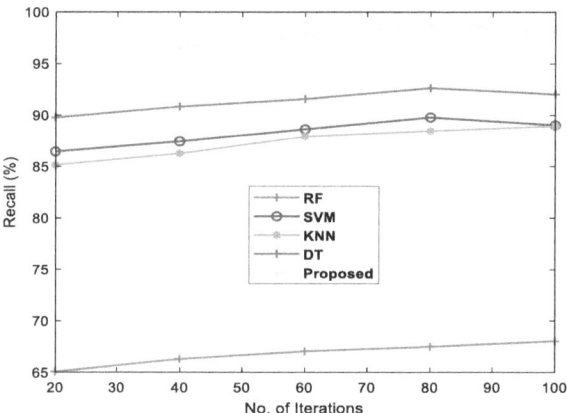

Fig. 3. Recall comparison graph

To have the option to introduce the exhibition of the proposed system, the highlights proposed in this work ($Angle, \widehat{dist}, \Delta dist_{R \to S}$) are secure in every conceivable element mixes. After a number of simulations, the combinations that typically perform better were chosen. The precision for the proposed DSVM with stacking is 98.27%. Hence the proposed is 5.66% better than the existing DT, 9.55% better than KNN, 8.46% better than SVM and 2.9% better than RF. This is pictorially illustrated in Fig. 2.

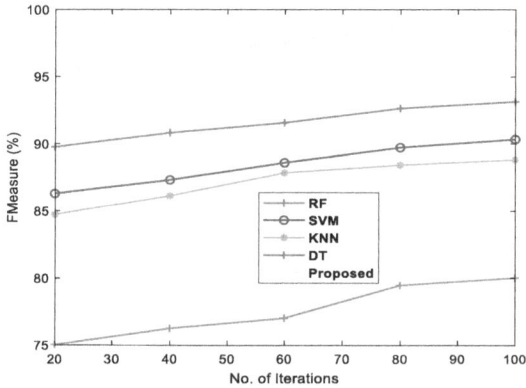

Fig. 4. F-Measure Comparison Graph

The recall for the proposed DSVM with stacking is 98%. Hence the proposed is 6.52% better than the existing DT, 10.23% better than KNN, 10.11% better than SVM and 44.12% better than RF. This is pictorially illustrated in Fig. 3.

The F-Measure for the proposed DSVM with stacking is 98.34%. Hence the proposed is 5.62% better than the existing DT, 10.74% better than KNN, 8.9% better than SVM and 22.92% better than RF. This is pictorially illustrated in Fig. 4. The accuracy for the proposed DSVM with stacking is 99.32%. Hence the proposed is 6.33% better than the existing DT, 10.6% better than KNN, 9.38% better than SVM and 25.4% better than RF. This is pictorially illustrated in Fig. 5.

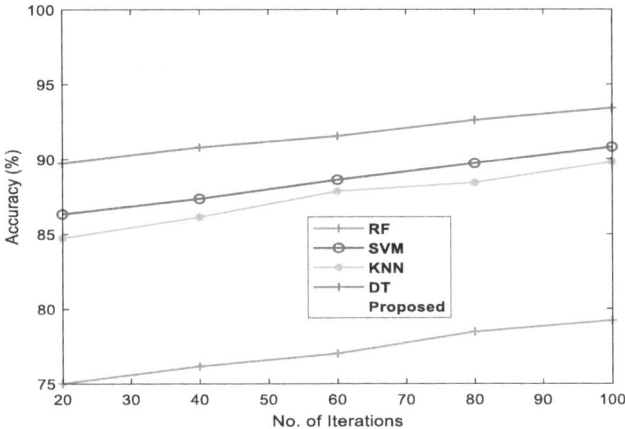

Fig. 5. Accuracy Comparison Graph

The computation time for the proposed DSVM with stacking is 38.5. Hence the proposed is better than the existing DT with 39 s, better than KNN with 40.1 s, better than SVM with 42 s and better than RF with 43 s. This is pictorially illustrated in Fig. 6.

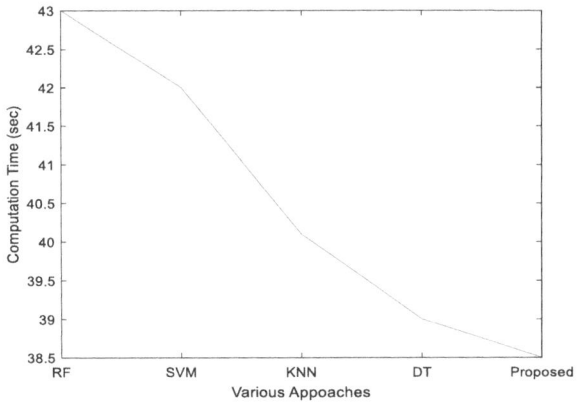

Fig. 6. Computation Time Comparison Graph

Hence from the comparison of the proposed with some existing approaches like RF, SVM, KNN and DT it is proved that the proposed model is providing improved results. Thus, with this algorithm the intrusion in VANET can be detected.

5 Conclusion and Future Work

For VANETs to detect position falsification attacks, a hybrid ML approach for IDS, including some suggested features, is presented in this paper. The notable ML methods, SVM and DT, and a group advancing by Stacking of SVM and DT are executed to fabricate an intelligent instrument. In order to distribute the attackers, the IDS scheme is applied to the vehicles. Every situation for various aggressor rates, assault as well as traffic densities kinds are investigated independently utilizing these three recognition strategies. VANETs are developed through this in-depth analysis for position falsification attacks in generalized discovery mechanism. The AoA among the sender as well as receiver, the evaluated distance estimated with the RSSI, along with the difference among the declared along with evaluated distance among the sender as well as receiver are all suggested features for detecting misbehavior. Further, this paper can be reached out to break down the continuous observing of quickly expanding network traffic.

References

1. Bangui, H., Buhnova, B.: Recent advances in machine-learning driven intrusion detection in transportation: Surv. Procedia Comput. Sci. **184**, 877–886 (2021)
2. Naqvi, I., Chaudhary, A., Kumar, A.: A systematic review of the intrusion detection techniques in VANETS. TEM J. **11**(2), 900 (2022)
3. Aboelfottoh, A.A., Azer, M.A.: Intrusion detection in VANETs and ACVs using deep learning. In: 2022 2nd International Mobile, Intelligent, and Ubiquitous Computing Conference (MIUCC), pp. 241–245. IEEE (2022)
4. Kadam, N., Raja Sekhar, K.: Machine learning approach of hybrid KSVN algorithm to detect DDoS attack in VANET. Int. J. Adv. Comput. Sci. Appl. **12**(7) (2021)
5. Wang, R.: Intrusion detection technology of Internet of vehicles based on deep learning. In: Huang, C., Chan, Y.W., Yen, N. (eds) 2020 International Conference on Data Processing Techniques and Applications for Cyber-Physical Systems. AISC, vol. 1379, pp. 323–329. Springer, Singapore (2021). https://doi.org/10.1007/978-981-16-1726-3_40
6. Kaur, G., Kakkar, D.: Hybrid optimization enabled trust-based secure routing with deep learning-based attack detection in VANET. Ad Hoc Netw. **136**, 102961 (2022)
7. Jeyaram, G., Madheswaran, M.: A secured Zobrist hash symmetric sentinel list based malicious attack detection in VANET. IETE J. Res. **69**(2), 611–622 (2022)
8. Singh, G., Khare, N.: A survey of intrusion detection from the perspective of intrusion datasets and machine learning techniques. Int. J. Comput. Appl. **44**(7), 659–669 (2022)
9. Haydari, A., Yilmaz, Y.: RSU-based online intrusion detection and mitigation for VANET. Sensors **22**(19), 7612 (2022)
10. Ercan, S., Ayaida, M., Messai, N.: Misbehavior detection for position falsification attacks in VANETs using machine learning. IEEE Access **10**, 1893–1904 (2021)
11. van der Heijden, R.W., Lukaseder, T., Kargl, F.: VeReMi: a dataset for comparable evaluation of misbehavior detection in VANETs. In: Beyah, R., Chang, B., Li, Y., Zhu, S. (eds) SecureComm 2018. LNICST, vol. 254, pp. 318–337. Springer, Cham (2018). https://doi.org/10.1007/978-3-030-01701-9_18

12. Bangui, H., Ge, M., Buhnova, B.: A hybrid data-driven model for intrusion detection in VANET. Procedia Comput. Sci. **184**, 516–523 (2021)
13. Karthiga, B., Durairaj, D., Nawaz, N., Venkatasamy, T.K., Ramasamy, G., Hariharasudan, A.: Intelligent intrusion detection system for VANET using machine learning and deep learning approaches. Wirel. Commun. Mob. Comput. (2022)
14. Alsarhan, A., Al-Ghuwairi, A.-R., Almalkawi, I.T., Alauthman, M., Al-Dubai, A.: Machine learning-driven optimization for intrusion detection in smart vehicular networks. Wirel. Pers. Commun. **117**, 3129–3152 (2021)
15. Vidhya, V., Madheswaran, M.: Walsh scalar Miyaguchi Preneel cryptography for secure image compression and storage in WANET. IETE J. Res. **69**, 1, 33–45 (2021, 2023)
16. Arya, M., et al.: Intruder detection in VANET data streams using federated learning for smart city environments. Electronics **12**(4), 894 (2023)
17. Ghaleb, F., et al.: Misbehavior-aware on-demand collaborative intrusion detection system using distributed ensemble learning for VANET. Electronics **9**(9), 1411 (2020)
18. Marwah, G.P.K., et al.: An improved machine learning model with hybrid technique in VANET for robust communication. Mathematics **10**(21), 4030 (2022)

Covid-19 Detection Using AI Deep Modified Resnet Model from Human Chest X-ray Images

Narenthira Kumar Appavu(✉) and Nelson Kennedy Babu

Saveetha Institute of Medical and Technical Sciences, Chennai, Tamil Nadu, India
{Narenthirakumara1025.sse,nelsonc.sse}@saveetha.com

Abstract. This research focuses on analyzing novel coronavirus (covid-19) from chest X-ray images with a deep learning (DL) - based model. The development of a revolutionary DL approach, Covid-19, is unique in this study of the development of a better residual Res-Net to create the proposed advanced Res-Net, Standard Res-Net 101 is required for tuning. The updated Res-net analyzed the new database made of 5,935 X-ray films, which were extracted from two database available in public. By reclaiming, multiplying and testing of many eras, our recommended model is considerably better than normal, healthy lung restrictions in identifying Covid-19 pneumonia. Rating measurements include memory, accuracy, recall, and F1 score and classification accuracy. Our suggested redemption continues to surpass other methods in the multiclassification problem using a pneumonia, lung-infused pulmonary and covid-19-infected lung samples. Detection Returns for Discrimination of COVID-19 in the test estimate for our advanced Res-net to use Resnet-101 as its foundation are 99.16%, 93.34%, and 92.71%for pneumonia and healthy normal lungs. The accuracy marks of our model for pneumonia and healthy normal lungs are 84.75% also 83.98%, respectively, which uses the ResNet-152 fine tune, respectively. Test results point out the possible use of our innovative CNN - determined model for classify the pneumonia and covid-19. The outcomes of this numerical study unequivocally demonstrate that these are outperforming the results of the prior study.

Keywords: Res-Net · X-ray · Enhanced Res-Net · Covid-19 · coronavirus · deep learning

1 Introduction

This could be an important step in the Covid-19 prevention plan. The unique coronavirus disease resulted in a worldwide epidemic (COVID-19) [1]. The bulk of the planet has been affected by it as of December 2019. Several typical indications, with as a cough, fever, a sore throat, shortness of breath and pneumonia are also present [2]. Lung infection or pneumonia is one of the common symptoms [3]. A chest X-ray will show it. Thus, early identification is essential to stop the coronavirus from spreading [4]. This could be an important step in the Covid-19 prevention plan [5]. Several countries have recently been affected by a second coronavirus outbreak. Because of the second wave, the daily infected rate and death rate are increasing. Early identification and management of COVID-19 is now defined in several ways.

There are three different diagnostic methods including viral testing, clinical imaging and blood tests. A blood test has the capability to distinguish antibodies to the SARS-CoV-2 coronavirus. Coronavirus antigens can also be found through viral testing. Inverse Copy Polymerase Chain Reaction, also known as RT-PCR, is a often used technique used for finding Covid-19. Unfortunately, it takings time to pinpoint the coronavirus. Also, multiple studies show that this test's sensitivity ranges from 50 to 62% [6]. This shows that the results of RT-PCR might not be trustworthy. Many RT-PCR as says are carried out to verify the accuracy of these test results [7].

Because RT-PCR is not widely available, making the diagnosis may be expensive for patients or health authorities in some countries [8]. Chest X-rays otherwise CT scans are utilised to diagnose the coronavirus, much like how X-rays are frequently used in radiological imaging to diagnosis various lung problems including pneumonia or other viral or bacterial infection. It can only show whether certain bacterial and viral diseases have any effects. According to the authors of [9], they were able to identify COVID-19 by CT scan, whereas RT-PCR came out negative for COVID-19. After that, the study investigated the compassion of CT-scan imageries and RT-PCR in patients with Covid-19 infection.

Even before symptoms occur in the early stages of coronavirus infection, abnormalities in X-rays and CT scans can be seen [10]. Images from a CT scan were used to locate COVID-19 [11]. In spite of this assistance, Testing facilities hardly ever use CT-scan pictures in clinics because the equipment is hard to come by and expensive. X-ray technology, which are frequently utilized to assess numerous ailments in underdeveloped nations. Computers and technology [12] have supported efforts to combat the COVID-19 epidemic. The detection and diagnosis of numerous diseases, including as stomach cancer, breast cancer, brain tumors, lung cancer, and spinal cord cancer, have been accomplished using deep learning (DL) techniques.

2 Related Works

In Covid-19, a sizable number of articles on digital radiography and X-rays were presented for diagnostic or therapeutic purposes. Certain noteworthy trainings will be emphasised in this. In a current study, a model called CovidX-NET was created [13]. Several deep learning (DL) models for binary classification, as well as Res_NetV2, DensNet_201, VGG-19, Inception_V3, MobileNet_V2, Xception, and Inception ResNet_V2, are compared in this study. They employed 50 samples of X-ray data, 25 of which were made up of Covid-19 and 25 of usual people [14]. The successful results of the tests are for two COVID-19 chest X-ray databases. It [15] and included data from 123 lung anterior view X-ray scans. The results were compared employing the ResNet50 and VGG19 models, respectively, is the Covid-Net model [16]. Work [17] used ImageNet previously trained using the Adam optimizer Normal and Covid-19 cases. Their accuracy rate of 93.3% was higher than that of earlier research. 13,975 X-ray pictures attained from various sources were cast-off [18]. The dataset remained retrieved in this study using four sources SIRM [19] and Radiopedia [20]. 700 images for pneumonia, 504 for normal, and 224 for COVID-19, respectively. In the proposed work literature review of pre trained model with dataset described and test dataset is then divided into training and test sections. When they used 50 X-rays for the experiment, the chest X-rays used were compiled from two different datasets, including Kaagle and Cohen [22]. Suggested by the authors of darknet [23], the 17 layers in the darknet model connect to the leaked ReLu. For multi classes, the accuracy of this model is 82.02%. Darknet achieved 98.08% accuracy for binary classification [24]. In them modified UNET model, the authors of [25] published their results with and without segmentation. Also, they used around 18480 chest X-ray images, which is around 3620 were of Covid-19 patient role. They obtained the photos they be situated considering from two different databases. Next divided the lungs into three categories: COVID-19-affected, opaque, and healthy [26]. In spite of these advantages, these models have some X-ray imaging limitations for the diagnosis of Covid-19. In light of this, we recommend Enhanced ResNet methods.

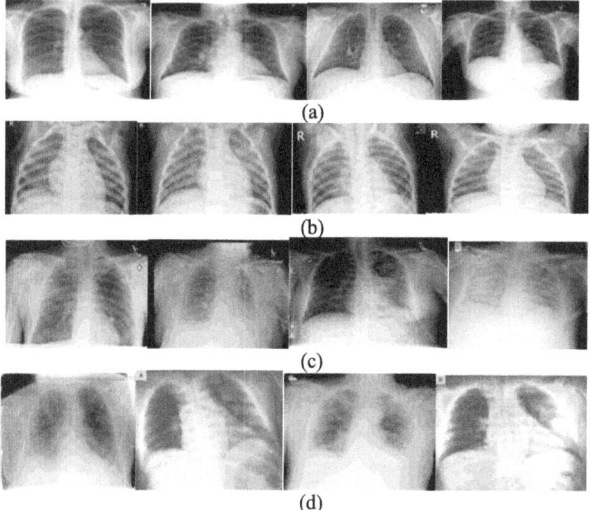

Fig. 1. Sample image of the (a) normal_x-ray, (b) pneumonia_x-ray, and (c) Lung_Opacity_x-ray (d) covid - X-ray images [21].

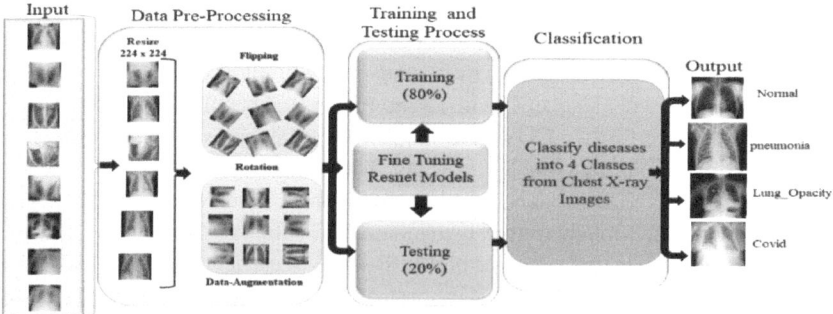

Fig. 2. Methodology of proposed Enhanced ResNet method.

Table 1. Proposed works Training and Testing Dataset details.

Number of Classes	Total_Set	Training Set (80%)	Testing set (20%)
Pneumonia	1345	1076	269
Normal	21169	16935	4234
Lung Opacity	6012	4809	1202
Covid	3616	2893	723

3 Materials and Methods

Two independent open-access sources served as the foundation for experimental data collection. From the of database of [27], chest-X-ray images healthy then pneumonia patients were extracted. There are around 4,275 pneumonia affected images and 1585 images of normal. After collecting these images, we obtained 79 chest X-rays from another dataset of patients sick with COVID-19 [28]. There was a total of 5,930 X-ray pictures. The dataset was then split into 20% training set and testing groups. Our test dataset is summarized in Table 1, and a sample dataset is shown in Fig. 1.

4 Methodology and Proposed Architecture

The entire structure of our proposed system is shown in Fig. 2. Our tests have gone through several stages.

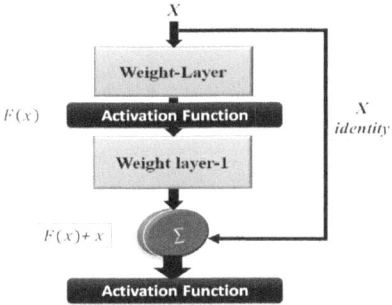

Fig. 3. ResNet architecture [29].

In addition to overfitting, X-ray image normalization and scaling have been performed to aid data generalization. Image resizing 224 × 224. Data augmentation is then applied We used our Enhanced ResNet model to train our data. After that to train our data for testing up to 35 epochs. Within 30 epochs, we achieved the highest accurateness for our Enhanced ResNet model. After that, we adjusted our model using hyperparameters. We evaluated the system as a whole using several performance measures.

4.1 Proposed Enhanced Res-Net

Enhanced Res-Net is used in this subsection to demonstrate the Res-Net paradigm. Figure 3 shows a neural network with a structure similar to Res-Net using x as input. Let's think about basic mappings and what we want to learn about $F(x)$. The Resnet's (activation function) ReLu is at its peak. A mapping is taken into account before the activation process. Consider the residual mapping in terms of $F(x)$ as well as the mapping. $F(x) + x$. Also, (4) shows the ResNet50 design. After that, Enhanced Res-Net (5) is generated using VGG's 3 by 3 convolutional algorithm. Res-Net's residual module 3 × 3 and 3 × 3 convolutional neural networks have comparable output.

Our proposed network architecture shown in Fig. 4. Which illustrate the enhanced model of proposed one. Each transform layer and ReLU function includes a block normalization. Enhansed ResNet CNN layer's primary components are transformation layers, an execution unit, pooling, block normalisation, etc. first The fundamental function of the convolution layer in neural networks is to extract particular information from input images.

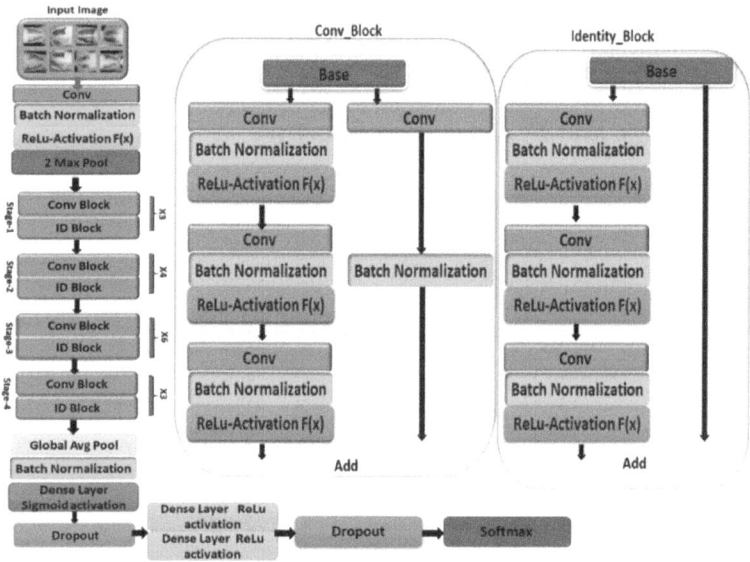

Fig. 4. Our proposed Enhanced ResNet architecture.

Filters are used one after the other to create this cycle. In this layer the input photos are used to generate feature maps. The productivity of each convolution layer is formerly fed into a convolution neural network. By doing this, non-linearity is incorporated into the structure. The ReLU operation is a well-known implementation method. Compared to other activation functions, this ReLU has lower computational cost and better gradient convergence. ReLU produces 0 if the input is negative (-ve) and the input is equivalent to the output if the input is positive. A pooling layer is created using the feature maps produced by the transformation procedures. This layer decreases the number of parameters measured thru the training process.

Also, it guarantees a fast computation time. Also, this layer helps streamline the overfitting process. In a maximum-pooling scenario, the maximum value of the input element is equal to the output. The output, however, is the average value of the input parts when mean-pooling is used. Block regularization testing is intended to improve the training period's integration quality. This layer formalizes the results of the previous layer. The ability to use a high learning rate is a benefit of this layer. Experimental results. Experiments and results are described in this section.

Colab, short for Google Collaboratory, conducted the experiments. Using the K80 GPU, Colab can guarantee a service (Tesla). 12 GB Nvidia guarantees 12 h of operation. If 10 more epochs flop to reduce the corroboration loss, the training phase is over. Next Epochs were set to 35 for a block size of 64 for experiments. The representations were accomplished for about 16 min. A multi-class classification was added as a result. As per a loss function, the categorical cross-entropy remained used. The output layer of their system exposes the SoftMax implementation. This can be described by the (1) below:

$$Ls = \sum_{o=1}^{L} \sum_{p=1}^{k} z_{op} 1_p q_{op} \qquad (1)$$

L and k denote the number of chest X-ray classes and the number of chest X-ray samples respectively denote Loss .z_{op} indicates that the o-th X-ray sample corresponds to the p-th X-ray class. The chest X-ray model's output o for class p is denoted by .q_{op}. Data augmentation, image normalization, First think about data pre-processing. This pre-processing improves the visual training function. Many elements can increase visual acuity. As portion of the pre-processing in this study, Next the photographs are converted to 224 × 224 pixels and their intensity is regularized. Intensity Normalization uses a "min-max normalization" approach to normalize the intensities of image pixels after their unique 0–255 values to a normal delivery.

This eliminates biased components and results in a uniform distribution. The union of the Enhanced ResNet procedure can be accelerated by this integration. Data Augmentation: The future Enhanced ResNet prototypical uses transfer learning, as a process of applying acquired knowledge. One model to another (base model) to form the target model. Transfer learning activities come in two different categories. Initially, static features are detached and then the prototypical is accomplished using data from the in-between layer. The second step is to adjust the data models. Finally, the fully connected (FC) layer is replaced by Enhanced ResNet, which applies ResNet 50's net weight, followed by ResNet 101, and then ResNet 152. Also, regularization and thick layers are taken into account. Figure 5 and Fig. 6 shown the proposed model for ResNet50. Enhanced ResNet accuracy and loss with train accuracy, value accuracy. The results of ResNet50 are shown in Table 2. For Covid-19, ResNet50 provides 99.34% accuracy. Also, for ResNet50, the accuracies of pneumonia and healthy lung were 89.76%, 89.12% and 99.34% respectively. The complete accurateness of this model is 92.74%.

The evaluation data for Enhanced ResNet is shown in Table 3, where the correct discovery rate for COVID-19 is 99.65%. Also, as indicated in Table 3, pneumonia-affected lungs and normal lungs have 92.78%, 92.23% and 99.65% accurate results,

Table 2. Routine Assessment of Resnet_101

Class	Correctness	Re-call	F1_Score
Normal	0.89	0.87	0.89
pneumonia	0.98	0.97	0,98
Covid-19	0.51	0.99	00.75

compared to ResNet152, as a result, the results of a healthy lung and later pneumonia are 81.43% and 81.15% correct.

Table 3. Performance Assessment of Res_net-152

Class	Correctness	Re-call	F1_Score
Normal	0.67	0.76	0.72
pneumonia	0.93	0.88	0.92
Covid-19	0.13	0.56	00.19

A straight evaluation is therefore not possible. Consequently, we contrasted our Enhanced Res-Net with ResNet-50 and ResNet-152. Figure 7 shows the heat map for pneumonia and Covid-19. On the leftward side of Fig. 7 the red hue of the heatmap is used to indicate the locations affected by pneumonia.

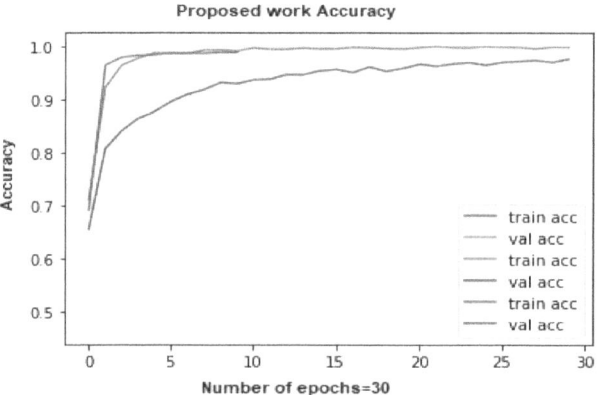

Fig. 5. Proposed model Accuracy.

The locations affected by Covid-19 are marked in red in the heatmap on the right side of Fig. 7. Our Enhanced ResNet model processed a heat map that shows the actual results indicating the relevant affected areas.

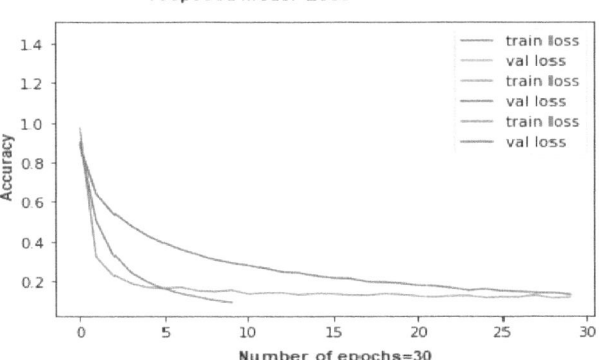

Fig. 6. Proposed model Loss.

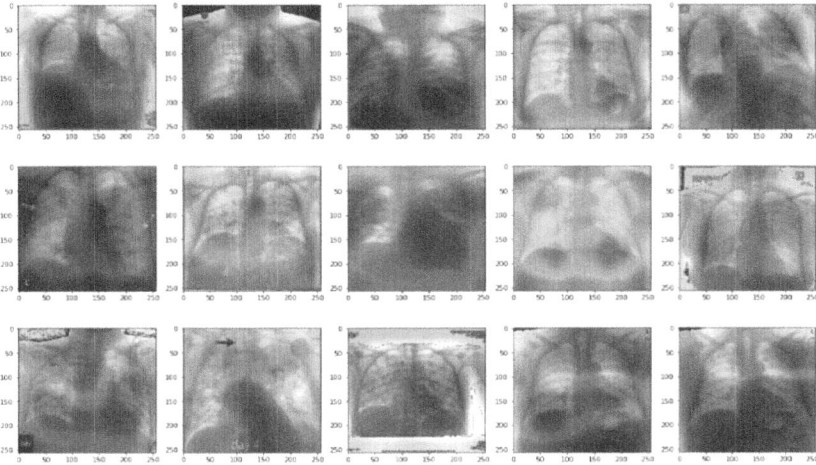

Fig. 7. Shows the heat map for pneumonia and Covid-19.

5 Conclusion

This paper proposes a unique Enhanced ResNet approach for multivariate classification of pulmonary infections including pneumonia and Covid-19. We compared the outcomes of normal, next pneumonia and finally covid19 infections. We created a dataset of 5,856 photographs from two different databases for testing and model evaluation purposes. The recommended Enhanced ResNet algorithm is ResNet101, improved by hyperparameter tweaking. Compared to other conventional Resnet models, it produced better results. The discovery rate of COVID-19 in the ResNet101 evaluation data was 98.74%. Furthermore, normal lungs and lungs damaged by pneumonia are identified with accuracy values of 93.34% and 92.71%, respectively. On the other hand, in the case of rehabilitation 152, the normal lungs damaged by pneumonia and lungs are identified with 84.75% and 83.98%

accurate levels respectively. Thermal diagram of the sites of damaged lung tissue was later displayed.

References

1. Chen, J.: Novel statistics predict the COVID-19 pandemic could terminate in 2022. J. Med. Virol. **94**(6), 2845–2848 (2022)
2. Fox, T., et al.: Antibody tests for identification of current and past infection with SARS-CoV-2. Cochrane Database Syst. Rev. (11) (2022)
3. Dinnes, J., et al.: Rapid, point-of-care antigen tests for diagnosis of SARS-CoV-2 infection. Cochrane Database Syst. Rev. (7) (2022)
4. Belkacem, A.N., Ouhbi, S., Lakas, A., Benkhelifa, E., Chen, C.: End-to-end AI-based point-of-care diagnosis system for classifying respiratory illnesses and early detection of COVID-19: a theoretical framework. Front. Med. **8**, 585578 (2021)
5. Kim, J.H., Marks, F., Clemens, J.D.: Looking beyond COVID-19 vaccine phase 3 trials. Nat. Med. **27**(2), 205–211 (2021)
6. Imran, A., et al.: AI4COVID-19: AI enabled preliminary diagnosis for COVID-19 from cough samples via an app. Inform. Med. Unlocked **20**, 100378 (2020)
7. Rutledge, R.G., Stewart, D.: Critical evaluation of methods used to determine amplification efficiency refutes the exponential character of real-time PCR. BMC Mol. Biol. **9**, 1–12 (2008)
8. C.-19 A.-U.-I.-I. Group: Early indicators of intensive care unit bed requirement during the COVID-19 epidemic: a retrospective study in Ile-de-France region, France. PLoS One **15**(11), e0241406 (2020)
9. Khatami, F., et al.: A meta-analysis of accuracy and sensitivity of chest CT and RT-PCR in COVID-19 diagnosis. Sci. Rep. **10**(1), 22402 (2020)
10. Das, D., Santosh, K.C., Pal, U.: Truncated inception net: COVID-19 outbreak screening using chest X-rays. Phys. Eng. Sci. Med. **43**, 915–925 (2020)
11. Shah, V., Keniya, R., Shridharani, A., Punjabi, M., Shah, J., Mehendale, N.: Diagnosis of COVID-19 using CT scan images and deep learning techniques. Emerg. Radiol. **28**, 497–505 (2021)
12. Sharma, H., Jain, J.S., Bansal, P., Gupta, S.: Feature extraction and classification of chest X-ray images using CNN to detect pneumonia. In: 2020 10th International Conference on Cloud Computing, Data Science & Engineering (Confluence), pp. 227–231 (2020)
13. Chamseddine, E., Mansouri, N., Soui, M., Abed, M.: Handling class imbalance in COVID-19 chest X-ray images classification: using SMOTE and weighted loss. Appl. Soft Comput. **129**, 109588 (2022)
14. Shibly, K.H., Dey, S.K., Islam, M.T.-U., Rahman, M.M.: COVID faster R-CNN: a novel framework to diagnose novel coronavirus disease (COVID-19) in X-Ray images. Inform. Med. Unlocked **20**, 100405 (2020)
15. Cohen, J.P., Morrison, P., Dao, L.: COVID-19 image data collection. arXiv Prepr. arXiv2003.11597 (2020)
16. Awan, M.J., Bilal, M.H., Yasin, A., Nobanee, H., Khan, N.S., Zain, A.M.: Detection of COVID-19 in chest X-ray images: a big data enabled deep learning approach. Int. J. Environ. Res. Public Health **18**(19), 10147 (2021)
17. Mahmud, T., Rahman, M.A., Fattah, S.A.: CovXNet: a multi-dilation convolutional neural network for automatic COVID-19 and other pneumonia detection from chest X-ray images with transferable multi-receptive feature optimization. Comput. Biol. Med. **122**, 103869 (2020)
18. Khan, I.U., Aslam, N.: A deep-learning-based framework for automated diagnosis of COVID-19 using X-ray images. Information **11**(9), 419 (2020)

19. Shuja, J., Alanazi, E., Alasmary, W., Alashaikh, A.: COVID-19 open source data sets: a comprehensive survey. Appl. Intell. **51**, 1296–1325 (2021)
20. I.S. Radiology, of M and I. Italian society of medical and interventional radiology (2020)
21. Bharati, S., Podder, P., Mondal, M.R.H.: X-ray images three levels. Figshare 2021 (2021). https://figshare.com/articles/dataset/X-ray_images_three_levels/14755965/1. Accessed 25 July 2021
22. Ghaleb, M.S., Ebied, H.M., Shedeed, H.A., Tolba, M.F.: COVID-19 X-rays model detection using convolution neural network. In: Proceedings of the International Conference on Artificial Intelligence and Computer Vision (AICV2021), pp. 3–11 (2021)
23. Alahmari, S.S., Altazi, B., Hwang, J., Hawkins, S., Salem, T.: A comprehensive review of deep learning-based methods for COVID-19 detection using chest X-ray images. IEEE Access (2022)
24. Ozturk, T., Talo, M., Yildirim, E.A., Baloglu, U.B., Yildirim, O., Acharya, U.R.: Automated detection of COVID-19 cases using deep neural networks with X-ray images. Comput. Biol. Med. **121**, 103792 (2020)
25. Shoaib, M.R., Elshamy, M.R., Taha, T.E., El-Fishawy, A.S., Abd El-Samie, F.E.: Efficient deep learning models for brain tumor detection with segmentation and data augmentation techniques. Concurr. Comput. Pract. Exp. **34**(21), e7031 (2022)
26. Mishra, M., Parashar, V., Shimpi, R.: Development and evaluation of an AI System for early detection of Covid-19 pneumonia using X-ray (Student Consortium). In: 2020 IEEE Sixth International Conference on Multimedia Big Data (BigMM), pp. 292–296 (2020)
27. Sethy, P.K., Behera, S.K.: Detection of coronavirus disease (Covid-19) based on deep features (2020)
28. Vinod, D.N., Prabaharan, S.R.S.: COVID-19-the role of artificial intelligence, machine learning, and deep learning: a newfangled. Arch. Comput. Methods Eng., 1–16 (2023)
29. Chen, Y., Qin, X., Wang, J., Yu, C., Gao, W.: FedHealth: a federated transfer learning framework for wearable healthcare. IEEE Intell. Syst. **35**(4), 83–93 (2020). https://doi.org/10.1109/MIS.2020.2988604

A Blockchain with RB-BM23-1 Method Used to Secure the Analyzed Data

Kishore, S. Rajaprakash[(✉)] [iD], K. Karthik, Neha, Sarangakrishna, and Pavan Chandra

CSE, Aarupadai Veedu Institute of Technology, Vinayaka Missions Research Foundation (DU), Chennai, Tamil Nadu, India
srajaprakash_04@yahoo.com, karthik@avit.ac.in

Abstract. A blockchain technology is very popular in the public sector in the "world-wide". Now-a-days every one interested to know about the blockchain technology history through online social networks such as Facebook or Twitter. In this paper, focusing on the history of blockchain and applying new security RB-BM23-1 method under in the blockchain technology. Proposed method is to find the combination of result in n-matrices. This method provide the good security with blockchain technology when compared to existing method ChaCha.

Keywords: Facebook · Twitter · Blockchain · ChaCha · RB-BM23-1

1 Introduction

A blockchain technology is very popular in the public sector in the "world-wide". Now-a-days every one interested to know about the blockchain technology history. A blockchain technology was proposed by "David Chaum" in the year of 1982; then later established chain of block security of cryptographically by "Stuart Haber and W. Scott Stornetta" in the year of 1991; then design the "Merkle Trees" for collecting the all documents into one block in the year of "1992–1993"; then "Bit Coin Gold" designed by "Nicholas Szabo" in the year of 1998.

Then laterally, the improve the bit coin design of block using "timestamp" and this method introduced in public sector for transaction of bitcoin by "Satoshi Nakamoto" in the year of 2008; the bitcoin allow to used people in blockchain technology in the year of 2013; the size of file is 20 GB for the bitcoin by used public and transaction of the records in blockchain technology in the year of 2014; and the next year size of the file is 30 GB.

The bitcoin file size was increased every day because of people's used bitcoin privately, and the year of 2016–2017 has 50 GB-100 GB of file size; the bitcoin "ledger size is exceed" 200 GB in the year of 2020 early; now bitcoin price is very high, so the second generation bitcoin introduced as "altcoin" in the year of 2020 shown in Fig. 1.

Concept of Blockchain: The concept of blockchain is consist of the many techniques and elements like cryptography, mathematics, algorithms, Peer-to-Peer network (P2P), distributed the algorithms, and economic model for solving the distributed the synchronise database problems.

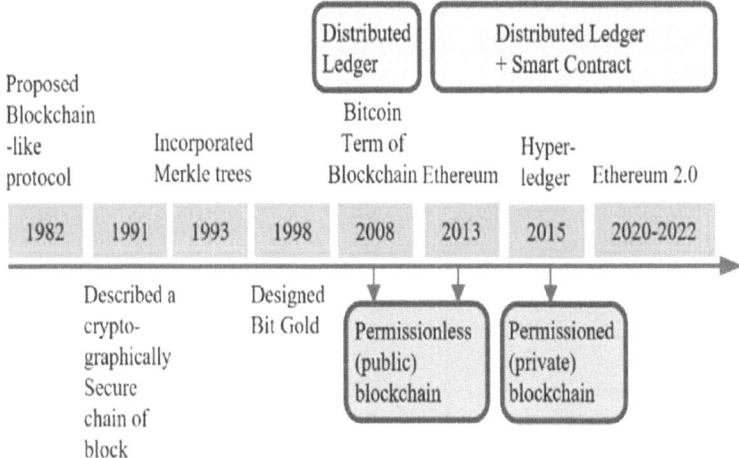

Fig. 1. Blockchain History

A blockchain technology has six main elements. 1. Decentralized: It store the data distributed way. 2. Transparent: All the user records will be stored transparently. 3. Open Source: Blockchain allowing to create the record and check the record publicly. 4. Autonomy: Every user device is update the data and transfer the data very safely. 5. Immutable: All the user records cannot be changed easily. 6. Anonymity: All the user transactions are high security.

Structure of Blockchain: A blockchain structure created by "Stuart Haber and W. Scott Stornetta" in the year of 1991. A blockchain structure is a decentralized and distributed the digital records to the public as blocks shown in Fig. 2.

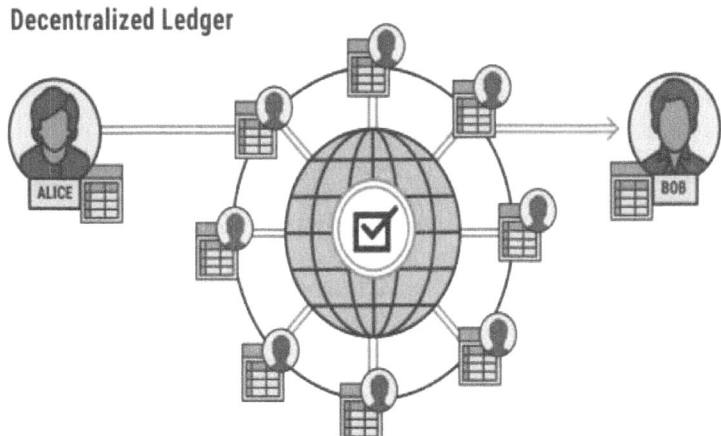

Fig. 2. A blockchain decentralized system

A blocks are used to transaction of the records is independently and that record cannot be modify and delete in the blocks because of each blocks must be connected to previous blocks are shown in Fig. 3.

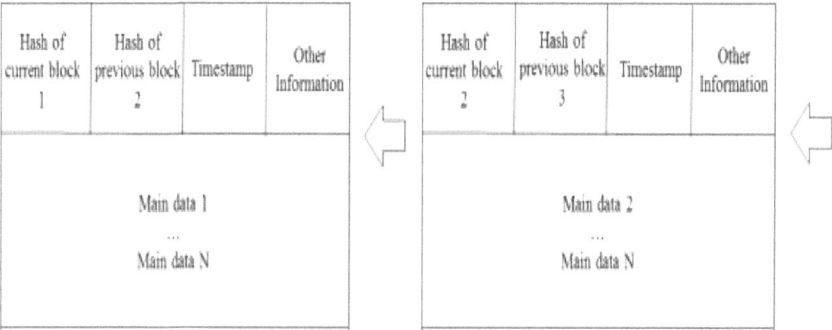

Fig. 3. Structure of blockchain

A blocks are store the information in hash and encoded the transaction by "Merkle Tree". Each transaction of the blocks has time. This block used to will add new security method as RajaprakashBagath-BlockMethod23-1 (RB-BM23-1).

2 Related Work

Vincent G is analyzing the bitcoin using blockchain algorithms [1]. Q. Zhang, Y. Li and R. Wang et al. are analyzed the attack and provide the security in blockchain using "decision bilinear Diffie–Hellman" [2]. Sarah Bouraga is analyzing the blockchain protocols survey [3]. T. McGhin et al. are discussed about the opportunities of the health care in blockchain [4]. A. Ghosh et al. are discussing about the challenges of the crypto currency for security in blockchain [5]. Huaqun Guo at.al are analyzing the history and applications of blockchain [6].

Hakima Khelifi et al. are focusing the performance improvement and security challenges in data networks [7]. Divyakant Meva is discussing the issues of the blockchain and development of the blockchain [8]. S. Ahmadjee et al. are discussing about the decisions of the threat and security in blockchain [9]. Erjon Hasanaj is talking about the security of the business industry in blockchain [10]. Minhaj Ahmad Khan et al. are talking about the survey of the security in IoT [11].

Satpal Singh Kushwaha et al. are discussing about the several directions security attacks in blockchain [12]. Author analyzed the data from Twitter through Machine Learning Algorithm [13].

They examined first Twitter information, then investigated anticipated that information by AI calculation, and applied SRB18 security calculation to that information, then contrasted the running time and Salsa for encryption [14]. They broke down the large information and stored that information with the insurance process applied for the RBJ20 calculation. This calculation has four phases used to safeguard the information [15]. They proposed 7 stages for providing the security of the data [16].

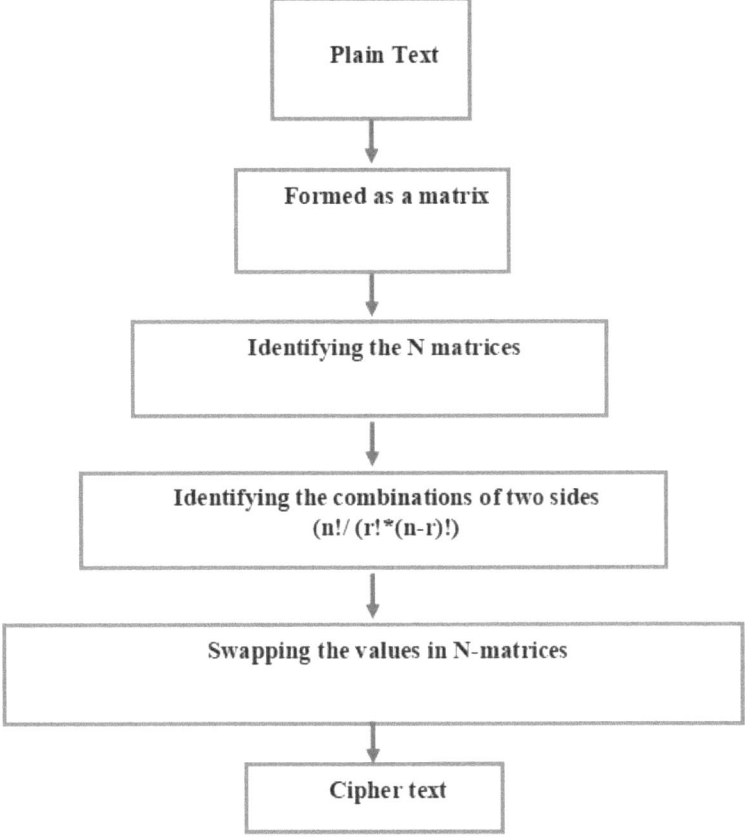

Fig. 4. RB-BM23-1 Methodology

3 Methodology

The proposed RB-BM23-1 method has three stages; 1st stage is identifying the n-matrices and 2nd stage is to find the combinations of two sides for given n-matrices using (n!/(r!*(n − r)!). The 3rd stage is to applying the result of combination in given n-matrices for swapping the values in Fig. 4.

Encryption Method.

- To create the two matrices A & B.
- To find the combination of two sides for Matrices A & B using (n!/(r!*(n − r)!).
- To merge the two matrices A & B as AB matrix.
- To merge the combinations of the three rows result as one row.
- To make a n-pair of combination result and used to swap the values in AB matrix.

4 Result

- Example for 2 matrices A & B.

$$A = \begin{bmatrix} A1 & A2 & A3 \\ A4 & A5 & A6 \\ A7 & A8 & A9 \end{bmatrix}$$

$$B = \begin{bmatrix} B9 & B8 & B7 \\ B6 & B5 & B4 \\ B3 & B2 & B1 \end{bmatrix}$$

- Combination of two possible numbers = (n!/(r!*(n − r)!) in Table 1.
- The 1st row combination result is 915327, the 2nd row combination result is 1511531, and 3rd row combination result is 170980.
- To merge the combination result and pair it (91), (53), (27), (15), (11), (53), (11), (70), (98), (00).

Table 1. Combination of 2 matrices

First ith Cells of A	1	2	3	Total No. Of Micro States
First jth Cells of B	9	6	3	
Probability of Micro States of First Cells	9	15	3	27
Second ith Cells of A	4	5	6	Total No. Of Micro States
Second jth Cells of B	6	5	4	
Probability of Micro States of Second Cells	15	1	15	31
Third ith Cells of A	7	8	9	Total No. Of Micro States
Third jth Cells of B	7	4	1	
Probability of Micro States of Third Cells	1	70	9	80

$$AB = \begin{bmatrix} A1B9 & A2B8 & A3B7 \\ A4B6 & A5B5 & A6B4 \\ A7B3 & A8B2 & A9B1 \end{bmatrix}$$

- The 1st pair swap value is 91

$$AB = \begin{bmatrix} A9B1 & A2B8 & A3B7 \\ A4B6 & A5B5 & A6B4 \\ A7B3 & A8B2 & A1B9 \end{bmatrix}$$

- The 2nd pair swap value is 53.

$$AB = \begin{bmatrix} A9B1 & A2B8 & A5B5 \\ A4B6 & A3B7 & A6B4 \\ A7B3 & A8B2 & A1B9 \end{bmatrix}$$

- The 3rd pair swap value is 27

$$AB = \begin{bmatrix} A9B1 & A7B3 & A5B5 \\ A4B6 & A3B7 & A6B4 \\ A2B8 & A8B2 & A1B9 \end{bmatrix}$$

- The 4th pair swap value is 15

$$AB = \begin{bmatrix} A3B7 & A7B3 & A5B5 \\ A4B6 & A9B1 & A6B4 \\ A2B8 & A8B2 & A1B9 \end{bmatrix}$$

- The 5th pair swap value is 11

$$AB = \begin{bmatrix} A3B7 & A7B3 & A5B5 \\ A4B6 & A9B1 & A6B4 \\ A2B8 & A8B2 & A1B9 \end{bmatrix}$$

- The 6th pair swap value is 53.

$$AB = \begin{bmatrix} A3B7 & A7B3 & A9B1 \\ A4B6 & A5B5 & A6B4 \\ A2B8 & A8B2 & A1B9 \end{bmatrix}$$

- The 7th pair swap value is 11

$$AB = \begin{bmatrix} A3B7 & A7B3 & A9B1 \\ A4B6 & A5B5 & A6B4 \\ A2B8 & A8B2 & A1B9 \end{bmatrix}$$

- The 8th pair swap value is 70.

$$AB = \begin{bmatrix} A2B8 & A7B3 & A9B1 \\ A4B6 & A5B5 & A6B4 \\ A3B7 & A8B2 & A1B9 \end{bmatrix}$$

- The 9th pair swap value is 98

$$AB = \begin{bmatrix} A2B8 & A7B3 & A9B1 \\ A4B6 & A5B5 & A6B4 \\ A3B7 & A1B9 & A8B2 \end{bmatrix}$$

- The 10th pair swap value is 00 (Table 2).

$$AB = \begin{bmatrix} A2B8 & A7B3 & A9B1 \\ A4B6 & A5B5 & A6B4 \\ A3B7 & A1B9 & A8B2 \end{bmatrix}$$

Table 2. RB-BM23-1 encryption performance

File Size	ChaCha	RB-BM23-1
14	2.1	2.6
66	2.4	2.9
302	2.8	3.5
812	3.1	4.1
1521	3.3	4.5
6570	3.9	4.9

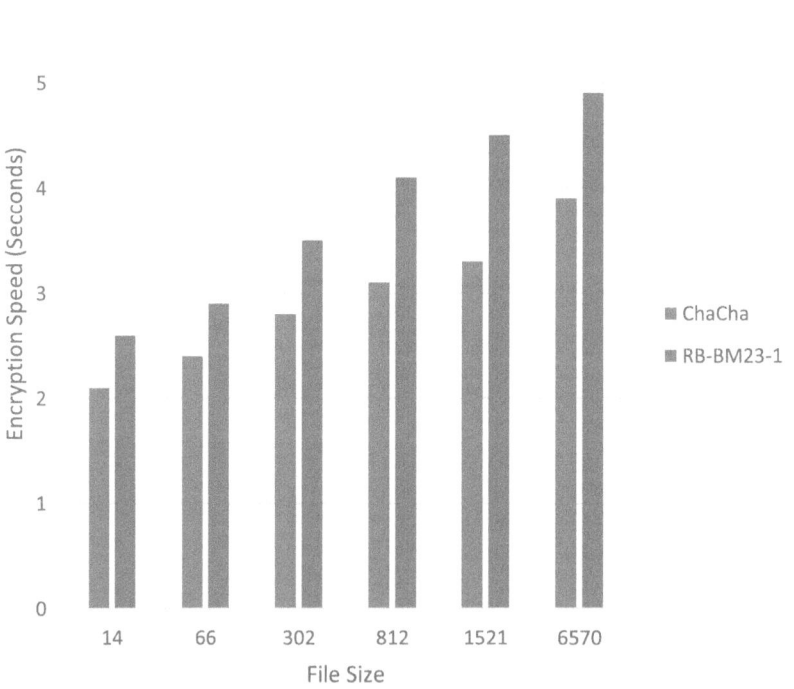

Fig. 5. Encryption performance

The proposed algorithm RB-BM23-1 compare the performance with existing method "ChaCha". The existing method is to do the process for move all diagonal values into the first column. The three by three matrix has (14, 66, 302, 812, 1521, 6570) bytes = > (6 × 6, 10 × 10, 15 × 15, 20 × 20, 40 × 40) matrix as shown in the Table 1.

From Fig. 5, the RP-BM23-1 method has compared the encryption speed in seconds. The encryption performance 2.6 (s), 2.9 (s), 3.5 (s), 4.1 (s), 4.5 (s) and 4.9 (s) for the RB-BM23-1. The RB-BM23-1 provide "more protection" of the data; when compared to existing method.

5 Conclusion

The author attempted to compile as many issues and challenges related to blockchain technology as possible. It has not been demonstrated that every security problem in ChaCha method. The proposed RB-BM23-1 method has three stages; 1st stage is identifying the n-matrices and 2nd stage is to find the combinations of two sides for given n-matrices using $(n!/(r!*(n-r)!)$. The 3rd stage is to applying the result of combination in given n-matrices for swapping the values. This method provide the good security with blockchain technology when compared to existing method ChaCha.

References

1. Gramoli, V.: From blockchain consensus back to Byzantine consensus. Future Gener. Comput. Syst., 760–769 (2020)
2. Zhang, Q., et al.: Blockchain-based asymmetric group key agreement protocol for internet of vehicles. Comput. Electr. Eng. (2020)
3. Bouraga, S.: A taxonomy of blockchain consensus protocols: a survey and classification framework. Expert Syst. Appl. (2021)
4. McGhin, T., Choo, K.-K.R., Liu, C.Z., He, D.: Blockchain in healthcare applications: research challenges and opportunities. J. Netw. Comput. Appl., 62–75 (2019)
5. Ghosh, A., Gupta, S., Dua, A., Kumar, N.: Security of cryptocurrencies in blockchain technology: state-of-art, challenges and future prospects. J. Netw. Comput. Appl. (2020)
6. Guo, H., Yu, X.: A survey on blockchain technology and its security. Blockchain Res. Appl. (2022)
7. Khelifi, H., Luo, S., Nour, B., Shah, S.C.: Security and privacy issues in vehicular named data networks: an overview. Mob. Inf. Syst. (2018)
8. Meva, D.: Issues and challenges with blockchain: a survey. Int. J. Comput. Sci. Eng., 488–491 (2018)
9. Ahmadjee, S., Mera-Gómez, C., Bahsoon, R., Kazman, R.: A study on blockchain architecture design decisions and their security attacks and threats. ACM Trans. Softw. Eng. (2022)
10. Hasanaj, E.: Blockchain and its security issues and challenges. Sci. Res. Pap. (2021)
11. Khan, M.A., Salah, K.: IoT security: review, blockchain solutions, and open challenges. Future Gener. Comput. Syst. (2017)
12. Kushwaha, S.S., Joshi, S., Singh, D., Kaur, M., Lee, H.-N.: Systematic review of security vulnerabilities in ethereum blockchain smart contract. IEEE Access (2021)
13. Bagath Basha, C., Somasundaram, K.: A comparative study of Twitter sentiment analysis using machine learning algorithms in big data. Int. J. Recent Technol. Eng., 591–599 (2019)

14. Bagath Basha, C., Rajapraksh, S.: Enhancing the security using SRB18 method of embedding computing. Microprocess. Microsyst. (2020)
15. Rajaprakash, S., et al.: RBJ20 cryptography algorithm for securing big data communication using wireless networks. In: Nagar, A.K., Jat, D.S., Marín-Raventós, G., Mishra, D.K. (eds.) Intelligent Sustainable Systems. LNNS, vol. 334, pp. 499–507. Springer, Singapore (2022). https://doi.org/10.1007/978-981-16-6369-7_46
16. Bagath Basha, C., et al.: A novel security algorithm RPBB31 for securing the social media analyzed data using machine learning algorithms. Wirel. Pers. Commun. https://doi.org/10.21203/rs.3.rs-1860348/v1

Development of an Interactive Road Safety System Using a Smart Helmet for Bike Users to Avoid Bike Accidents

S. Rajalingam(✉), S. Kanagamalliga, K. Sakthi Priya, R. Karpaga Priya, and S. Kavitha

Saveetha Engineering College, Chennai, Tamil Nadu, India
rajalingams@saveetha.ac.in

Abstract. Two-wheelers have a higher accident rate than vehicles, trucks, and buses in modern life. In order to avoid two-wheeler accidents, there are many existing methodologies, which even failed to reduce the accident rate. One such system, which shall be improved, is the smart helmet system. The difficulties we encounter on the roadways every day in the actual world serve as the inspiration for this project. Road accidents are becoming more frequent day by day, and in nations like India, where bikes are more common, many people lose their lives due to negligence caused by not wearing helmets. Although helmets are widely accessible, few people wear them. The proposed system includes various vehicle sensors in the smart helmet with suitable algorithms. The two-wheeler cannot be started without a helmet and hence the rider's safety is guaranteed. If an accident does occur, our system will alert the ambulance about it so that they can take the necessary precautions to preserve the life of the accident victim. It was created with Arduino. We install sensors on the various sides of the helmet, and they are linked to an Arduino board. As a result, when the bike rider crashes, sensors detect it, and the Arduino uses the GPS that is connected to it to retrieve GPS location data. The GSM module automatically sends messages to ambulances, police, and family members when sensor data exceeds the pressure limit. The rider can use the SMS-sending stop switch to halt message transmission in the event of minor injury.

Keywords: Rider Safety · GSM · GPS · Road Accident

1 Introduction

Over 1,000 persons under the age of 25 passed away on the roads each day or roughly 400,000 people annually. This demonstrates the necessity to pay attention to raising young people's awareness of concerns related to road safety in order to prevent such losses [1]. It is most common for drivers between the ages of 16 and 24 to be involved in such accidents. Due to their lack of experience, the majority of children and teenagers misjudge danger or fail to recognise threats. Additionally, they are more prone to drive too fast and leave insufficient space between cars. Families should therefore be aware of their children's driving habits [2–4].

Also, there is a persistent increase in demand for two-wheelers in the market, which is driving up production and sales. Two-wheeler use will increase along with the economy, raising safety issues that must be addressed [5].

Motorcycle accidents most frequently result from negligence on the part of one or more parties. Moreover, two-wheelers are particularly susceptible to suffering injuries in collisions due to the lack of physical protection they have. As a result, there is an increasing need to upgrade motorbike safety features [6]. Also, parents' ongoing worries about their kids riding motorcycles need to be taken into account. Thus, as technology advances, it is necessary to use better techniques for ensuring safety and interfacing with the vehicle.

Vehicle occupants were 12 times more likely to die than cyclists. (72 vs. 6 fatalities per billion kilometres), according to the Federal Highway Administration and the Highway Traffic Safety Agency [7]. Cycling accidents can have serious repercussions, including serious injuries, fatalities, and economic loss owing to the traffic backups they generate. As a result, it becomes urgently important to address the research challenge of how to increase riding safety while reducing the risk and consequences of accidents that cyclists may experience [8]. They are becoming a significant issue for many transportation agencies in major cities throughout the world, which has significantly increased the amount of study done on bicycle safety [9].

Despite growing concerns, there is still a dearth of scientific risk analysis and safety management on cycling safety in the literature, especially when compared to other modes of transportation [10]. This paper develops a novel conceptual risk analysis method based on a Bayesian network in order to close this research gap. (BN). Using information obtained from six years' worth of transport accident reports involving riding in the Liverpool city region (2012–2018), this method will enable the analysis and prediction of the severity of cycling accidents[11].

The paper is structured as follows. Section 2 discusses studies on riding safety using risk assessment approaches and evaluates recent literature on bike accidents to identify associated hazards and risks. The methods and approaches used in this work are discussed in Sect. 3, which is followed by Sect. 4's results and discussion. The paper is concluded in Sect. 5 with a discussion of its contributions and consequences.

2 Existing Methodology

The current project has wireless communications that are linked to smartphones. The communication hardware is employed to immediately dial a predetermined emergency contact when this prototype detects a crash or accident [6, 12, 13]. The speed that the rider is travelling at can be controlled by a second device. The helmet is fixed with all the parts and sensors that measure the bike's speed and give the rider instructions on how to change their pace in response to impending impediments. In areas without traffic checks, riders do not wear helmets. In large nations like India, it is hard to test the amount of alcohol in each rider's blood. Traffic police have a difficult time enforcing the law [6, 14, 15].

Next, preliminary analyses of all the cycling-related road accident report in the Liverpool city region from 2012 to 2017 were conducted in order to determine the initial primary risk information. To support the primary cycling risk factors that have been identified, these original accident data are given [16]. The interdependence of the risk factors and their combined impact on accident severity is determined through statistical analysis and utilised as the input to a risk analysis and prediction model to support the development of bicycle safety policies. The BN model can be used to examine the key risk factors that have an impact on how serious bike accidents are [8, 17, 18].

Empirical cases based on the new set of data gathered from the incidents recorded in 2018 are also employed [19–21] in order to assess the BN model and its predicted accuracy in a variety of dangerous scenarios as well as to produce meaningful information for accident prevention. In order to decrease the severity of possible bicycle accidents, safety recommendations are made to riders and transit authorities based on the findings [22].

The following features of this research are novel: 1) The risk factors influencing the severity of bike accidents are found by combining relevant literature with actual historical accident statistics. 2) For the creation of a data-driven BN risk model, which contains more than 200,000 pieces of risk information on more than 100 different risk parameters, large data are gathered and processed. 3) By including sophisticated uncertainty modelling, it expands the corpus of knowledge on quantitative cycling risk evaluations. (e.g. BN). 4) In order to confirm the dependability and precision of the risk prediction model, a fresh set of data collected in 2018 is employed. The method, which can accurately predict risk intensity in over 95% of actual cases, can provide insightful information for formulating policies.

3 Proposed Methodology

The proposed method comprises of two modules. They are the main module and the bike module. In Figs. 1 and 2, the Block Diagram of the Proposed System is displayed. The first part is the front module where all the sensor setup and software and transmitter end are placed. Another part is the receiver part which is set up at bike inside the unit, to stop the vehicle once any of the features reaches an abnormal condition. In the below module part, it consists of the Sensor section, Transmitter section, Arduino Controller to control those components signal, Power supply, LCD Display and GPS. The module's alcohol sensors determine whether the rider is intoxicated enough to drive. When a rider is intoxicated, the motorbike is not allowed to start and the rider is not permitted to operate the vehicle. In this setup, an Arduino microcontroller is connected to an alcohol sensor, which continuously provides signals to the microcontroller while monitoring the user's breathing. When a sensor sends an alcohol signal, the microcontroller sends the information to the motor. Engine starting requires pressing a button on the system. The system locks the engine if alcohol is found. The system also uses GSM and GPS to transmit a message that says, "Accident occurred," along with the latitude and longitude of the occurrence. A vibration sensor is used to identify accidents. In order to prevent circuit damage, it also incorporates a temperature sensor that alerts the user when the helmet becomes too hot.

In the below module part, it consists of the Motor (To stop the ignition of the bike once it reaches an abnormal level), the Receiver section, the Arduino Controller to control those component's signal, Power supply.

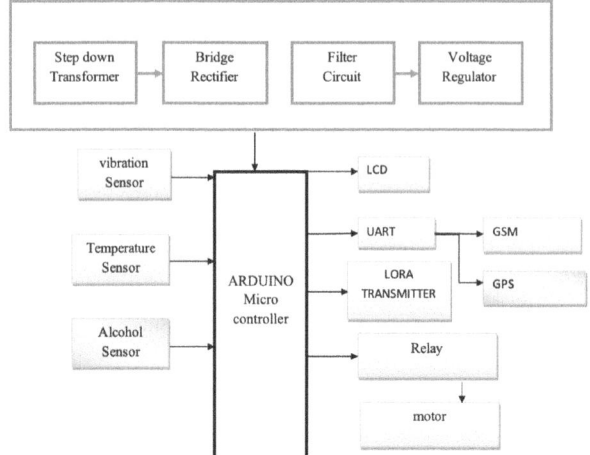

Fig. 1. Block diagram of Main Front Module

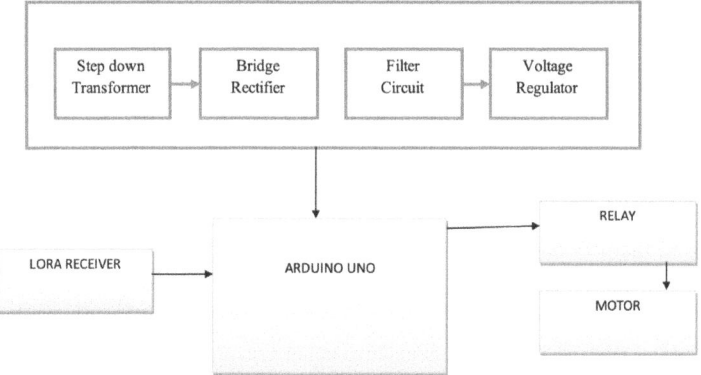

Fig. 2. Block diagram of Bike Unit module

3.1 Temperature Sensor

In this paper, the LM34 sensor was used. In industrial automation, the most often measured process variable is temperature. A temperature sensor is most frequently used to translate temperature values into electrical values. Temperature sensors are essential for accurate temperature readings and temperature control in industrial applications. There are several different kinds of temperature sensors. Sensors vary greatly in terms of characteristics including contact type, temperature range, calibration technique, and sensing element. The sensor element in the temperature gauges is housed in plastic or metal housings. The sensor will reflect changes in ambient temperature with the aid of conditioning circuits.

We employ the LM34 series of temperature sensors in the functional module we designed for measuring temperature. The output voltage of the precision integrated-circuit temperature sensors in the LM34 series is linearly proportional to the temperature in Fahrenheit. In comparison to linear temperature sensors calibrated in degrees Kelvin, the LM34 is advantageous since the user does not need to deduct a significant constant voltage from the output to obtain suitable Fahrenheit scaling. There is no need for external calibration with the LM34 Vibration Sensor.

We employed a tilt vibration sensor in this paper. There are many different projects, devices, and applications that use vibration sensors. The tool you're probably using is referred to as a vibration sensor, regardless of whether you're trying to estimate a vehicle's speed or the strength of an incoming earthquake. Some of them run on their own, while others need a separate power supply.

The necessity for a variety of sensors is caused by various machine working conditions relating to temperature extremes, magnetic fields, vibration range, frequency range, electromagnetic compatibility (EMC), electrostatic discharge (ESD) conditions, and the needed signal quality.

3.2 Alcohol Sensor

In this work, the MQ-X Gas Sensor is utilised. A device for determining the blood alcohol concentration (BAC) of a breath sample is known as a breathalyser or breathalyser (a combination of the words "breath" and "analyzer" or "analyser"). The brand name (a generalised trademark) for the device invented by Robert Frank Birkenstein to measure blood alcohol content is Breathalyzer. Although it was officially trademarked on May 13, 1954, the phrase is frequently used to describe any generic equipment for calculating blood alcohol levels. A colourless, flammable liquid with a faint chemical odour, ethanol is volatile and combustible. Due to its low freezing point, it is also employed as a fuel, disinfectant, solvent, and active fluid in many alcohol thermometers.

Being composed of two ethyl groups joined by hydroxyl groups, the structure is straightforward. Alcohol sensor to detect the presence of alcohol vapours in Breathalyzers or alarm systems. This sensor device has extremely high sensitivity and a quick response time. The device has great stability and long life and may be operated using a straightforward drive circuit. The equipment is prepared to analyse a new sample once all of the acetic acids have been removed from the fuel cell. Although it was officially trademarked on May 13, 1954, the phrase is frequently used to describe any generic equipment for estimating blood alcohol levels. A comprehensive alcohol sensor module for Arduino or SeeDuino is called Grove - Alcohol Sensor.

An alcohol sensor MQ303A semiconductor is used in its construction. It reacts to alcohol quickly and with good sensitivity. It can be used to create Breathalyzers. All the circuitry required by the MQ303A is implemented in this Grove, including power conditioning and a heater power supply. This sensor generates a voltage that is inversely proportional to the amount of alcohol in the air.

4 Results and Discussion

The three major parameters considered for the decision-making event are (i) the Lifespan of the engine, (ii) Accident detection and (iii) Alcohol consumption detection.

4.1 Lifespan of Engine

The temperature sensor detects the heat of the engine once the bike's engine has started, and it alerts the other end when the engine temperature reaches the reference value. (Depends on the vehicle). When the Temperature reaches the Value of 50 °C, The Engine of the Bike will be Off and the Emergency contact will receive the notification. When Bike Engine gets heated up or goes beyond the reference temperature, The temperature Sensor senses it and turns the ignition of the bike which led to stopping the bike. Once it reaches the maximum limit, the time to cool down the Bike engine is given and then it is ignited. After the cooling period of the engine, The Bike can be ignited (Get Started).

The emergency condition is notified when the temperature reaches 0 °C as shown in Fig. 3. Now the Tabular Column as in Table 1and Graphical representation of the Temperature (0 °C) with respect to the Time (Hr) and if the temperature reached the abnormal limit, it will notify the emergency contact, which is stored in the Software. The Sample message of it is shown in Fig. 4

Table 1. The lifespan of the engine

Time (hours)	Temperature (Celsius)
0	18
1	20
2	23
3	26
4	30
5	35
6	37
7	39
8	45
9	45
10	49
11	49
12	50
13	51

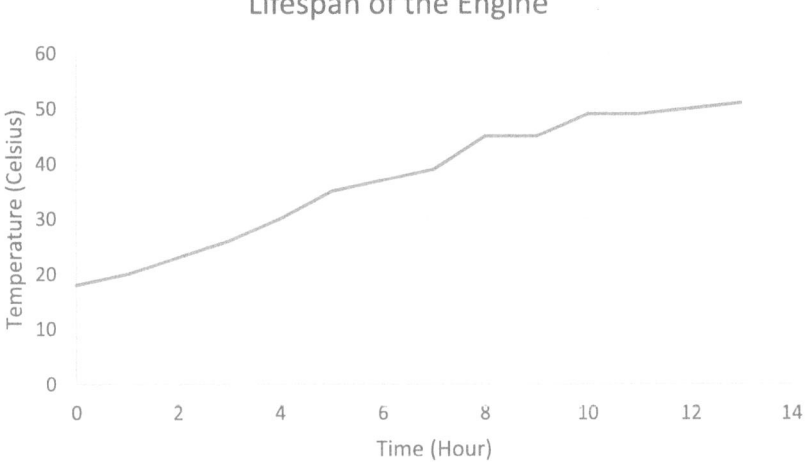

Fig. 3. Graphical Representation of Lifespan of Engine

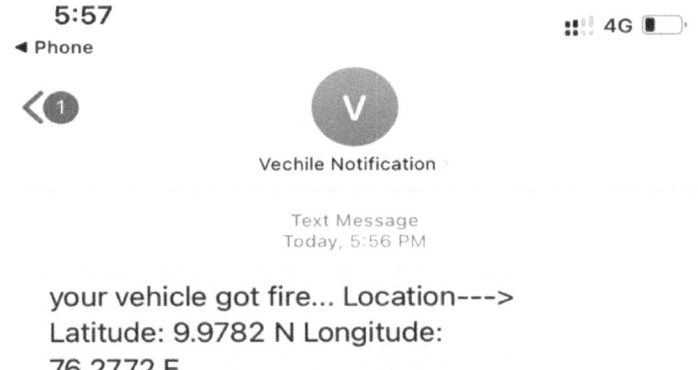

Fig. 4. Alert message1

4.2 Accident Detection

When Bike is met with an accident, the Vibration sensor placed in the vehicle will get activated and send the signal to the receiver via transmitter and stop the bike. The Bike is initially in moving mode or on standby. If any accident occurred the vehicle, the signal is sent to the receiver which stops the bike and notifies the other end person. When the vibration sensor is sensed, the signal is sent to stop the engine of the bike and notify the emergency contact person of the location of the bike via GPS.

In the below image, V is denoted as Vibration detection in the bike. The value is displayed as 1 when the bike met with the accident else it will remind as 0. Now the

Tabular Column as in Table 2 and Graphical representation of the Vibration (Hertz) with respect to the Time (Mins) and If the Vibration reached the abnormal limit, it will notify the emergency contact stored in the Software as shown in Fig. 5.

Table 2. Accident detection

Time (min)	Vibration (Hz)
1	0
2	15
3	10
4	34
5	15
6	67
7	70
8	56
9	45
10	20
11	1

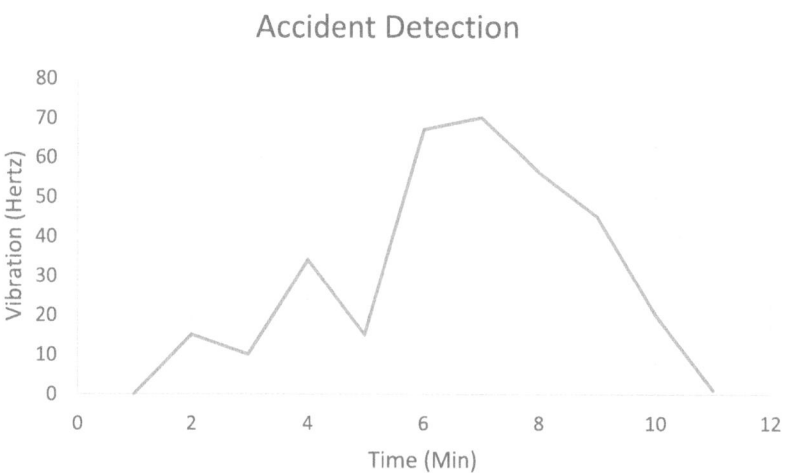

Fig. 5. Graphical Representation of Accident Detection

4.3 Alcohol Detection

When a biker drinks, the alcohol sensor or gas sensor detects the amount of alcohol in the blood and sends a signal to the receiver to shut off the bike's engine. Once the Bike is started, The Biker's alcohol level is tested via breathing of the biker by the alcohol sensor which is set up on the helmet of the biker. Once it reaches the maximum limit, it will send the signal to the receiver and stops the bike. Once the biker's blood alcohol content hits the reference limit, the alcohol sensor detects it. To stop the bike and alert the emergency contact, it will transmit a signal to the receiver. G represents the biker's alcohol consumption level in Fig. 6.

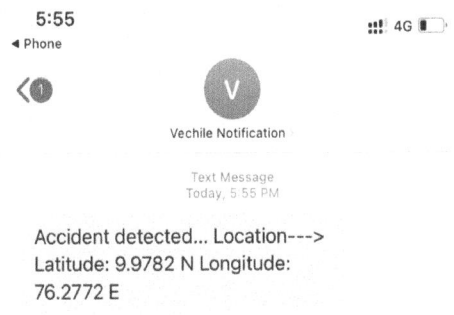

Fig. 6. Alert message2

Now the Tabular Column as in Table 3 and Graphical representation of the Alcohol Level with respect to the Time (Mins) is shown in Fig. 7 and If the level of alcohol consumption reached the abnormal limit, it will notify the emergency contact which is stored in the Software. The Sample message of it is shown Fig. 8.

Table 3. Alcohol detection

Time (min)	Alcohol level (cm)
1	100
2	234
3	456
4	479
5	657
6	738
7	788
8	839
9	900
10	922

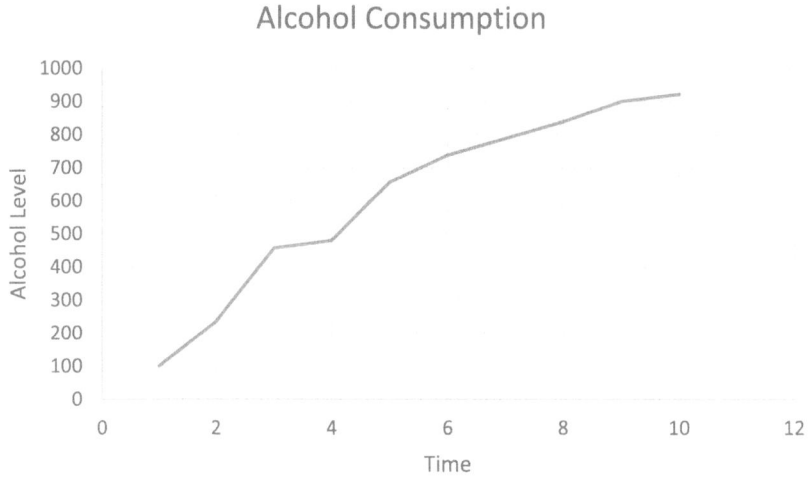

Fig. 7. Graphical Representation of Alcohol Level

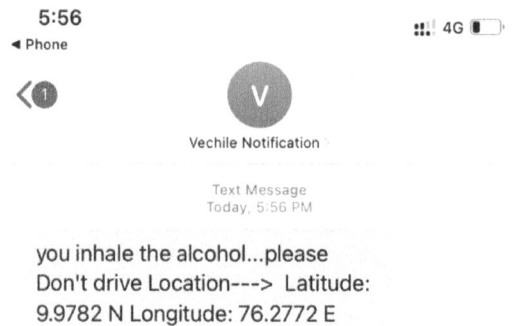

Fig. 8. Alert message3

5 Conclusion

The design and implementation of a bike safety system with experimental findings are presented in this paper in order to prevent accidents. The Accident with Two-wheeler can be avoided by this smart device. The most common Drink and Drive cases can be reduced with the incorporated technology. The Engine performance can be improvised by the proposed efficient method. The frequent notification of the status of the biker /vehicle helps the concerned person to track the vehicle status. The Lifetime of the Bike engine can be revised and maintained in the proper manner.

References

1. Burian, S.E., Liguori, A., Robinson, J.H.: Effects of alcohol on risk-taking during simulated driving. Hum. Psychopharmacol. Clin. Exp. **17**, 141–150 (2002)
2. Fu, L., Wang, J.X.: A vehicle on-board alcohol detection system based on integrated multi-sensors. Adv. Mater. Res. **490–495**, 91–94 (2012)
3. Li, F.P., Jiang, M.L., Guo, Z.Y.: Intelligent vehicle mounted alcohol detection control system. Appl. Mech. Mater. **109**, 248–252 (201)
4. Manzoni, V., Corti, A., Spelta, C., Savaresi, S.M.: A driver-to-infrastructure interaction system for motorcycles based on smartphone. In: 2010 13th International IEEE Annual Conference on Intelligent Transportation Systems, Madeira Island, Portugal, 19–22 September (2010)
5. Spelta, C., Manzoni, V., Corti, A., Goggi, A., Savaresi, S.M.: Smartphone-based vehicle-to-driver/environment interaction system for motorcycles. IEEE Embed. Syst. Lett. **2**(2), 39–42 (2010)
6. Wu, Y., Abdel-Aty, M., Zheng, O., Cai, Q., Yue, L.: Developing a crash warning system for the bike lane area at intersections with connected vehicle technology. Transp. Res. Rec. **2673**(4), 47–58 (2019)
7. Hamim, O.F., Hoque, M.S., McIlroy, R.C., Plant, K.L., Stanton, N.A.: A socio-technical approach to accident analysis in a low-income setting: using accimaps to guide road safety recommendations in Bangladesh. Saf. Sci. **124**, 104589 (2020)
8. Wismans, J., et al.: Implications of road safety in national productivity and human development in Asia. In: Eighth Regional EST Forum in Asia, Colombo, Sri Lanka. Chalmers University in Gothenburg (2014)
9. Mejia, D., Gomez, S., Martinez, F.: A low-cost wearable autonomous system for the protection of bicycle users. Int. J. Adv. Comput. Sci. Appl. **14**(1) (2023)
10. Pauzié, A., Gelau, C.: Methodology for safety and usability assessment of ITS for riders. In: Proceedings of the European Conference on Human-Centred Design for Intelligent Transport Systems (2010). Bundesanstalt für Straßenwesen.
11. Yang, Z., et al.: Risk analysis of bicycle accidents: a Bayesian approach. Reliab. Eng. Syst. Saf. **209**, 107460 (2021)
12. Kovaceva, J., Bärgman, J., Dozza, M.: On the importance of driver models for the development and assessment of active safety: a new collision warning system to make overtaking cyclists safer. Accid. Anal. Prev. **165**, 106513 (2022)
13. He, Y., Sun, C., Chang, F.: The road safety and risky behaviour analysis of delivery vehicle drivers in China. Accid. Anal. Prev. **184**, 107013 (2023)
14. Kapousizis, G., et al.: A review of state-of-the-art bicycle technologies affecting cycling safety: level of smartness and technology readiness. Transp. Rev., 1–23 (2022)
15. von Sawitzky, T., Grauschopf, T., Riener, A.: Hazard notifications for cyclists: comparison of awareness message modalities in a mixed reality study. In: 27th International Conference on Intelligent User Interfaces (2022)
16. Carter, D., et al.: Road Safety Fundamentals: Concepts, Strategies, and Practices that Reduce Fatalities and Injuries on the Road, No. FHWA-SA-18-003. United States. Federal Highway Administration. Office of Safety (2017)
17. Williams, B.: Bicycle crash detection: using a voice assistant for more accurate reporting. Diss. University of Oregon (2018)
18. Bhowmick, R., et al.: Cloud-based intelligent accident proof helmet and detect state of intoxication. In: 2023 Third International Conference on Artificial Intelligence and Smart Energy (ICAIS). IEEE (2023)
19. Wang, J., Chen, Y., Chen, D.: Analyzing safety concerns of (e-) bikes and cycling behaviours at intersections in the urban area. Sustainability **14**(7), 4231 (2022)

20. Xu, J.: From individual to connected driving: an interactive driver vehicle environment for road user safety. Diss. Northeastern University (2017)
21. Jost, G., et al.: A challenging start towards the EU 2020 road safety target. Eur. Transp. Saf. Counc. (2012)
22. Mohan, D., et al.: Road Traffic Injury Prevention Training Manual. World Health Organization (2006)

Customer Relationship Management a Decision Support System: Bibliometric Analysis 1990–2023

S. Lokesh[ID] and S. Vasantha[✉]

School of Management Studies, Technology and Advanced Studies (VISTAS), Vels Institute of Science, Chennai 600091, India
vsantha.sms@velsuniv.ac.in

Abstract. A decision support system (DSS) that incorporates customer relationship management (CRM) can help businesses make more informed decisions about their customers and improve their overall customer experience. CRM provides business insight to various levels of management to take day-to-day operation decisions to strategic management level decisions. The study aims to analyze 4,982 publications indexed in SCOPUS between 1990 and 2023 by keyword threshold in customer relationship management (CRM) and decision support systems (DSS). Bibliometric analysis is made for the science mapping technique by the VOSviewer application based on the number of documents per year, author, affiliation, country, citations, type, subject area, and keywords. The finding was that the year 1990 had the first publication in DSS; author Smith, A.D., defined 47 publications; Robert Morris University, with a higher affiliation, is located in Pennsylvania, United States of America, and the major 998 publications Computer Science had a total of 2339 publications, which was 27.7% more than other sectors. Article publications were indexed in 43.9% of publications. The future implications of CRM as a decision support system for operations are likely to involve greater personalization, multi-channel integration, increased use of data analytics, more automation, and increased collaboration. One potential novelty idea based on the findings of this study could be to develop a decision support system that incorporates machine learning and artificial intelligence to provide even more advanced insights into customer relationship management. This system could analyze customer data from multiple sources, including social media, website behavior, and purchase history, to provide personalized recommendations and insights for businesses. Additionally, the system could use natural language processing to analyze customer feedback and sentiment analysis to provide businesses with a better understanding of their customers' needs and preferences. This more advanced decision support system could help businesses stay ahead of the competition and provide an even better customer experience.

Keywords: Bibliometric Analysis · Customer Relationship Management (CRM) · Decision Support System (DSS)

1 Introduction

1.1 Customer Relationship Management

One of the main concepts in CRM research is how to use technology to improve interactions with customers. [1] found in their study that mobile CRM affects customer satisfaction and loyalty. The results show that mobile CRM technology has a positive effect on customer satisfaction and loyalty, especially when it comes to personalization and convenience.

[2] conducted a study on Cargo field from Indonesia found the customer loyalty was influenced by CRM and customer satisfaction. [3] looks at how social media can be used to improve customer engagement in the hospitality business. The results show that social media can be a good way to keep customers interested and build relationships that last. Finally, there is growing interest in the ethical dimensions of CRM. A study by [4] examines the role of ethics in CRM and argues that ethical considerations should be central to CRM strategies.

The authors propose a framework for ethical CRM that includes principles such as transparency, fairness, and respect for privacy. [5] reviewed the literature on the impact of technology on CRM. The author found that technology has greatly facilitated CRM implementation, as it enables companies to store and analyses vast amounts of customer data, which can be used to improve their products and services. Furthermore, technology has allowed companies to engage with their customers through various channels, such as social media, email, and Chabot's, which can enhance the customer experience.

[6] reviewed the literature on the role of social media in CRM. The authors found that social media can facilitate customer engagement and communication, which can improve the customer experience. [7] conducted an empirical investigation of CRM practices in the UK. They found that effective CRM implementation can lead to improved customer retention, satisfaction, and profitability. Furthermore, they found that CRM can enhance the customer experience by enabling companies to offer personalized products and services.

1.2 Decision Support System (DSS)

In today's fast-paced business environment, making decisions quickly and accurately is essential for organizations to stay competitive. Decision support systems (DSS) have emerged as effective tools to assist managers in making informed decisions based on relevant data. Many researchers have studied the role of DSS in improving decision-making processes in various industries. According to [8], DSS can be divided into four categories: data-driven DSS, model-driven DSS, knowledge-driven DSS, and communication-driven DSS. Data-driven DSS uses historical and current data to support decision-making. Model-driven DSS uses mathematical models to analyze data and provide insights.

Knowledge-driven DSS use expert knowledge to provide recommendations, while communication-driven DSS focus on collaboration among decision-makers. [9] looked at how DSS affects the performance of an organization's supply chain. Their study found that the use of DSS significantly improved operational and financial performance.

Similarly, [10] found that the use of DSS improved decision-making and resulted in cost savings for organizations. In addition to improving decision-making processes, DSS has also been shown to have a positive impact on organizational learning. According to [11] DSS can help organizations learn from past decisions and improve future decision-making.

DSS can also facilitate knowledge sharing and collaboration among decision-makers, leading to a more informed decision-making process. However, there are inevitable difficulties throughout the implementation of DSS. According to [12] issues such as resistance to change, lack of IT infrastructure, and the complexity of DSS can hinder the adoption and effectiveness of DSS in organizations. Therefore, it is important for organizations to carefully plan and implement DSS to ensure successful adoption and maximum benefits.

1.3 Integration of Customer Relationship Management and Decision Support System

The integration of CRM and DSS can provide businesses with a competitive edge by helping them make informed decisions based on customer data. According to [13], the integration of CRM and DSS can help businesses better understand their customers' needs, preferences, and behaviors. This understanding can enable businesses to develop personalized marketing strategies and improve customer satisfaction. Additionally, the integration of CRM and DSS can help businesses identify and target profitable customers while minimizing the costs of customer acquisition and retention. Argue that the integration of CRM and DSS can improve the quality of decision-making by providing decision-makers with accurate and timely information. DSS can provide businesses with real-time data about customer behavior, sales trends, and inventory levels, which can help businesses make informed decisions about pricing, product development, and inventory management. This information can also help businesses identify potential opportunities and threats in the market, allowing them to adapt quickly to changing market conditions.

1.4 Benefits of Customer Relationship Management and Decision Support System Integration

The integration of CRM and DSS can provide businesses with several benefits, including increased sales, improved customer satisfaction, and reduced costs. According to [14] the integration of CRM and DSS can help businesses improve their customer retention rates by providing them with personalized experiences that meet their unique needs and preferences. This, in turn, can lead to increased customer loyalty and repeat purchases, which can increase sales and profitability.

Additionally, the integration of CRM and DSS can help businesses to reduce their costs by minimizing the resources required for customer acquisition and retention. According to [15] the integration of CRM and DSS can help businesses identify their most profitable customers and focus their resources on retaining them. This can help businesses reduce their costs by avoiding unnecessary expenditures on less profitable customers.

The integration of CRM and DSS can provide businesses with a competitive edge by helping them make informed decisions based on customer data. The integration of CRM and DSS can improve the quality of decision-making, increase sales, improve customer satisfaction, and reduce costs [16]. Businesses that integrate CRM and DSS can improve their operations and increase their profitability by identifying and targeting profitable customers while minimizing the costs of customer acquisition and retention.

2 Methodology

Bibliometric studies are a form of quantitative analysis that focuses on the study of publications and their characteristics. It involves the use of bibliographic data to analyze patterns of authorship, publication, citation, and collaboration within a particular field of research.

By using specific keywords and limiting the search to certain language or document type parameters, researchers can identify a subset of publications that are relevant to their research question. This can help to streamline the data collection process and ensure that the analysis is focused on the most relevant and informative publications.

2.1 Search Strategy and Data Collection

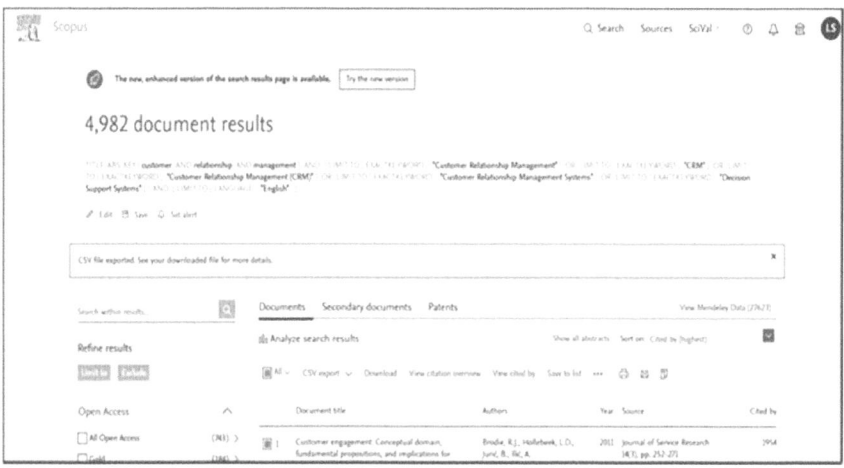

Fig. 1. Document search results of SCOPUS

As per Fig. 1. TITLE-ABS-KEY (customer AND relationship AND management) AND (LIMIT-TO (EXACTKEYWORD, "Customer Relationship Management") OR LIMIT-TO (EXACTKEYWORD, "CRM") OR LIMIT-TO (EXACTKEYWORD, "Customer Relationship Management (CRM)") OR LIMIT-TO (EXACTKEYWORD, "Customer Relationship Management Systems") OR LIMIT-TO (EXACTKEYWORD,

"Decision Support Systems")) AND (LIMIT-TO (LANGUAGE, "English")) 4,982 document results, year range 1990–2023 Query returned 4,982 document results and that the search was conducted on publications with a year range from 1990 to 2023. This information is important for understanding the scope of the study and the extent of the data collection process.

3 Results

3.1 Publications by year

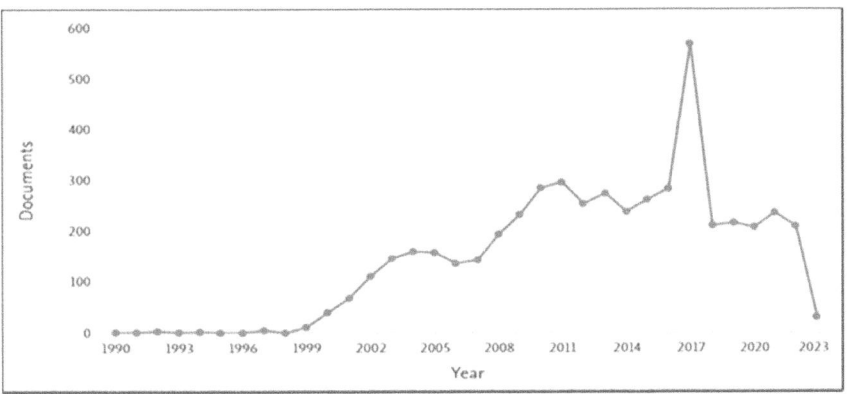

Fig. 2. Publications 1990–2023

Based Fig. 2 the list of years and publications, the number of publications has been increasing steadily since the early 2000s, with a peak in 2017 (568 publications). This could reflect a growing interest in research in these subject areas, or it could be due to changes in the way that research is conducted and disseminated. The number of publications has declined somewhat in recent years (2020–2022), although it is too early to say whether this represents a trend or a blip. There are relatively few publications from before the 2000s, which could be due to a number of factors, such as changes in the way that research is conducted and reported, or a lack of archival data. The number of publications varies quite a bit from year to year, with some years (e.g. 2017) having many more publications than others (e.g. 1999). This could be due to changes in funding, changes in the popularity of certain research topics, or other factors. The number of publications in 2023 (30) is relatively small compared to some of the earlier years on this list, but this is likely due to the fact that it is still early in the year and many publications may not have been reported yet. Overall, this list of years and publications provides some insight into the trends and fluctuations in research in these subject areas over the past few decades.

3.2 Publications by Author and Citations

The information provided in Fig. 3 and Table 1 is a list of authors with their respective number of publications and citations. The number of publications refers to the number of research papers or articles that each author has published, while the number of citations indicates the number of times that other researchers have cited their work in their own papers.According to the list, Lu J. has published only two articles but has received an impressive 795 citations. Neslin S.A. has published ten articles and received 279 citations, followed by Verhoef P.C. with 16 articles and 118 citations. Shankar V. has published only three articles but received 117 citations, while Hollebeek L.D. has published only two articles but has received 42 citations. Van den Poel D. has published the highest number of articles on the list, with 20 publications, but received only 24 citations. Zhang G. and Yen D.C. have published five and ten articles respectively but received only 24 and 19 citations each. Kumar V. has published 14 articles and received 11 citations, while Wang W. has published seven articles but received only seven citations.

It's important to note that this information is based on a specific dataset and may not reflect the authors' full body of work or impact in their respective fields.data shows the number of publications authored by each author in a bibliometric study. The data is sorted by Top 10 based on the number of publications. The author with the highest number of publications is Smith, A.D., with 47 publications, although Van Den Poel, D. is the second most prolific author with 20 publications, followed by verhoef p.c. with 16 publications. This information can provide insights into the most prominent authors in the field of customer relationship management. These authors may have made significant contributions to the field or have a strong influence on the direction of research. Additionally, this information can help to identify potential collaborators or experts in the field for future research projects.

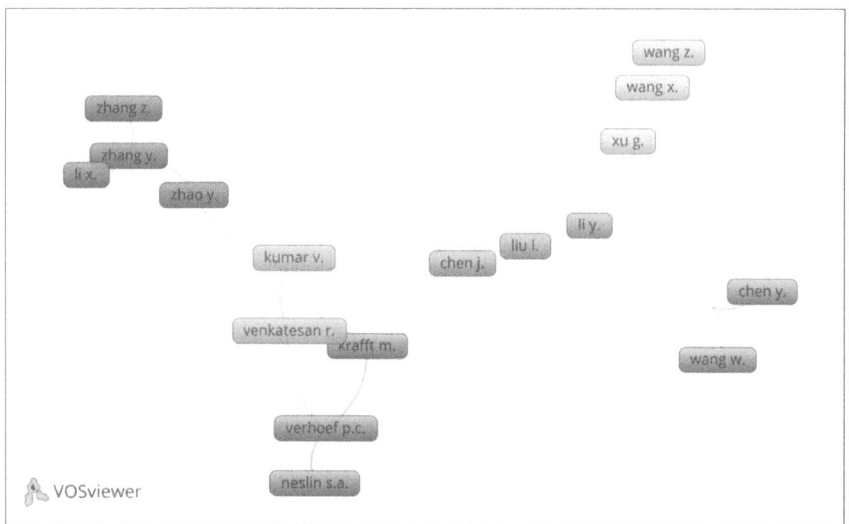

Fig. 3. Publications by Author

Table 1. Author's based on Publications and Citations

Sr. No	Author	Publications	Citations
1	Lu J	2	795
2	Neslin S.A	10	279
3	Verhoef P.C	16	118
4	Shankar V	3	117
5	Hollebeek L.D	2	42
6	Van Den Poel D	20	24
7	Zhang G	5	24
8	Yen D.C	10	19
9	Kumar V	14	11
10	Wang W	7	7

3.3 Publications by Type

Data as per Fig. 4 represents the document types of publications analyzed in a bibliometric study. The study analyzed a total of 4961 publications, with the majority being articles (2188), followed by conference papers (1862) and book chapters (626). Other document types included reviews (213), short surveys (25), notes (24), retracted articles (24), conference reviews (8), editorials (6), books (4), and letters (1). This information provides insights into the types of publications that are commonly analyzed in bibliometric studies. The presence of retracted articles is also notable, as it suggests that the study took steps to account for potentially problematic publications in its analysis.

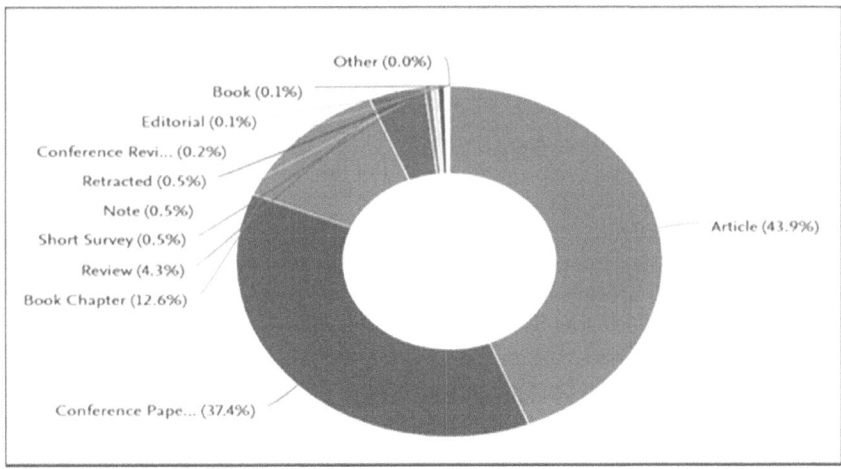

Fig. 4. Publications by Type

3.4 Publications by Affiliations

The number of publications by Fig. 5 is based on affiliated institute and the data is sorted by the number of publications in descending order. Robert Morris University has the highest number of publications with 47 affiliations, followed by Hong Kong Polytechnic University with 43 affiliations, and the University of Tehran with 34 affiliations. This information can provide insights into the institutions that are actively researching in the field of customer relationship management. Other factors such as citations, impact factor of the journals where their research is published, and the relevance of their contributions to the field should also be considered.

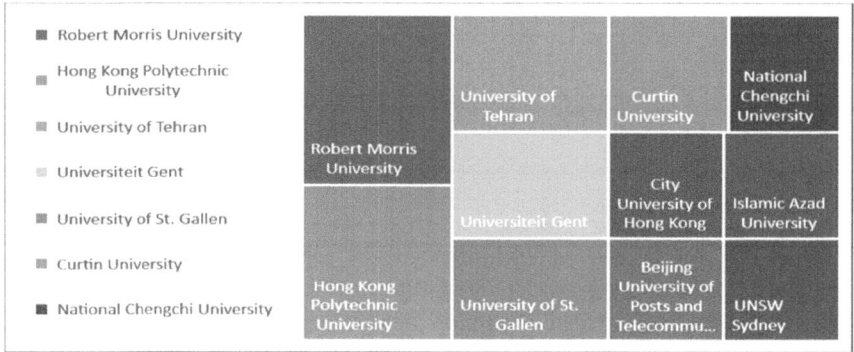

Fig. 5. Publications by Affiliations

3.5 Occurrence of Keywords

By the VOSviewer application, the result referring to Fig. 6 and Table 2 Keyword occurrences refer to the frequency with which a particular keyword appears in a given context or dataset. In this case, the context is a list of keywords and their corresponding frequencies. From the provided list, "Customer Relationship Management" appears most frequently with 1541 occurrences, followed by "Public Relations" with 673 occurrences and "Sales" with 546 occurrences. "Customer Satisfaction" has 407 occurrences, while the acronym "CRM" appears 365 times. "Data Mining" has 293 occurrences, and "Customer Relationship Management (CRM)" has 276 occurrences. "Marketing" appears 163 times, "Information Management" has 146 occurrences, "Decision Making" has 113 occurrences, and "Decision Support Systems" has 88 occurrences.

These keyword occurrences provide insight into the most frequently discussed topics in the areas of customer relationship management, public relations, sales, data mining, and information management. The high frequency of certain keywords such as "Customer Relationship Management" highlights the importance of this topic in various industries. These findings can be used to guide further research and decision-making in these areas.

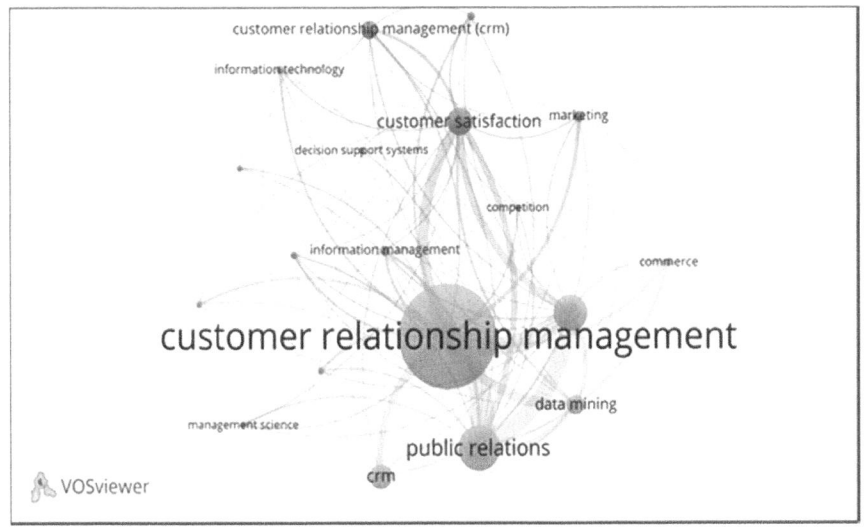

Fig. 6. Occurrence of Keywords

Table 2. Author's Keywords based on occurrence

Sr. No	Keyword	Occurrences
1	Customer Relationship Management	1541
2	Public Relations	673
3	Sales	546
4	Customer Satisfaction	407
5	CRM	365
6	Data Mining	293
7	Customer Relationship Management (CRM)	276
8	Marketing	163
9	Information Management	146
10	Decision Making	113
11	Decision Support Systems	88

3.6 Publications by Country and Citations

The Vosviewer processed the data as shown in Fig. 7. The United States has the highest number of documents (514) and citations (27357) Table 3 among the listed countries, followed by China with 183 documents and 4264 citations. The United Kingdom has the third-highest number of documents (153) and the fourth-highest number of citations (5949). Taiwan has the third-highest number of citations (4702), while India has the fourth-highest number of documents (154) and the fifth-highest number of citations

(4647). It's also worth noting that link strength is a measure of the quality of the links between the documents, with higher values indicating stronger connections. The United States has the highest link strength (221), followed by the United Kingdom (93) and Germany (73). Overall, the data suggests that the United States is a dominant player in the field of research, with a significant lead over other countries in terms of both document output and citation impact. However, China, Taiwan, and India are also making significant contributions to the field, as evidenced by their high citation counts.

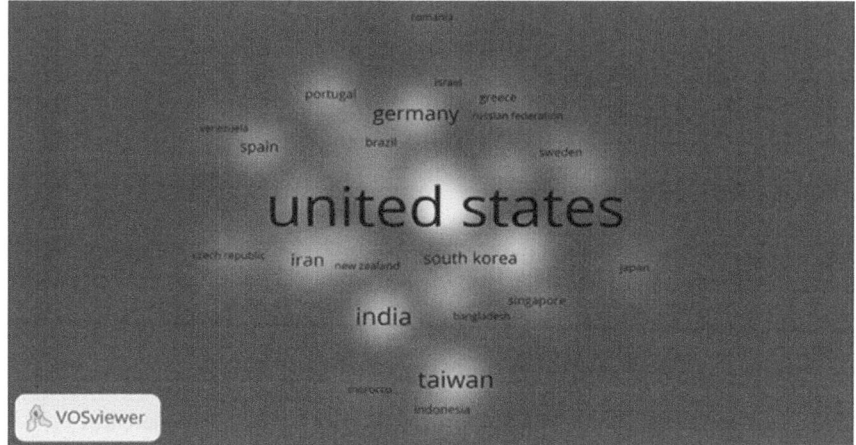

Fig. 7. Publication by Countries

Table 3. Publications by Country's based on Citation

Sr. No.	Country	Documents	Citations
1	United States	513	51
2	China	183	90
3	India	153	485
4	United Kingdom	153	18
5	Germany	117	46
6	Australia	92	8
7	Spain	60	12
8	Hong Kong	56	29
9	Canada	48	5
10	Netherlands	42	4

3.7 Sponsorship

Based on Fig. 8. The Top 10 sponsorship is listed out of total of 158 institute at 512 sponsorships. Natural Science Foundation of China has the highest number of publications with 83, followed by the national science foundation with 18 publications. The Fundamental Research Funds for the Central Universities, the European Commission, and Hong Kong Polytechnic University have similar numbers of publications with 16, 13, and 13, respectively. The other sponsors, including the National Science Council, Fundacao Para a Ciencia e a Tecnologia, National Council for Scientific and Technological Development (Conselho Nacional de Desenvolvimento Científico e Tecnológico), European Regional Development Fund, Ministry of Education of the People's Republic of China, and National Research Foundation of Korea, have contributed to a relatively smaller number of publications, with the number of publications ranging from 9 to 12.

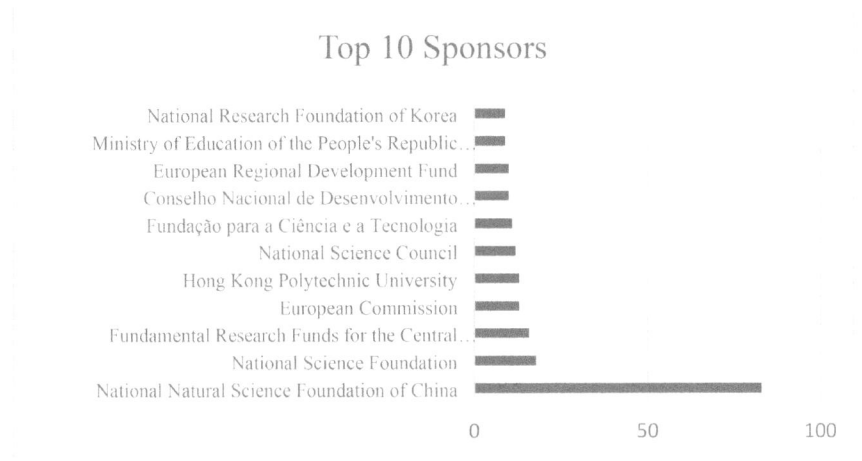

Fig. 8. Top 10 Sponsorships

It is important to note that the number of publications alone does not provide a complete picture of the sponsorship contribution. factors such as the quality and impact of the publications, the funding amount provided by the sponsors, and the research areas covered by the publications should also be considered when evaluating the sponsorship contribution.

4 Discussion and Conclusion

Query returned 4,982 document results and the search was conducted on publications with a year range from 1990 to 2023. Based on the list of years and publications, the number of publications has been increasing steadily since the early 2000s, with a peak in 2017 (568 publications). The number of publications has declined somewhat in recent years (2020–2022), although it is too early to say whether this represents a trend or a blip. The number of publications authored by each author in a bibliometric study is sorted by the Top 10 based on the number of publications. The author with the highest number of publications is Smith, A.D., with 47 publications.

Van Den Poel, D. is the second most prolific author with 19 publications, followed by Fae, A. with 17 publications. This information can provide insights into the most prominent authors in the field of customer relationship management, which may have made significant contributions to the field or have a strong influence on the direction of research. The study analyzed 4961 publications, with the majority being articles (2188), followed by conference papers (1862) and book chapters (626).

Other document types included reviews (213), short surveys (25), notes (24), retracted articles (24), conference reviews (8), editorials (6), books (4), and letters (1). The number of publications affiliated with each institution in a bibliometric study was sorted by the number of publications in descending order, with Robert Morris University having the highest number of publications with 47 affiliations, followed by Hong Kong Polytechnic University with 43 affiliations, and the University of Tehran with 34 affiliations. The frequency and link strength of these keywords provided insights into the topics discussed in the text and their relative importance. The United States had the highest number of documents (514) and citations (27357) among the listed countries, followed by China with 183 documents and 4264 citations. The United Kingdom has the highest number of documents (153) and the fourth-highest number of citations (5949).

Taiwan has the third-highest number of citations (4702), while India has the fourth-highest number of documents (154) and the fifth-highest number of citations (4647). The United States has the highest link strength (221), followed by the United Kingdom (93) and Germany (73). Overall, the United States is a dominant player in the field of research, with a significant lead over other countries in terms of both document output and citation impact. China, Taiwan, and India are also making significant contributions to the field, as evidenced by their high citation counts. The Top 10 sponsorship is listed out of 158 institutes at 512 sponsorships.

The National Natural Science Foundation of China has the highest number of publications with 83 publications, followed by the National Science Foundation with 18 publications. Other sponsors, such as the National Science Council,Conselho Nacional de Desenvolvimento Científico.

This bibliometric study provides valuable insights into the field of customer relationship management, including the most prominent authors, institutions, and countries contributing to research. While the number of publications has fluctuated over time, it is clear that the United States remains a dominant player in the field, with China, Taiwan, and India also making significant contributions. By analyzing the frequency and link strength of keywords, this study sheds light on the topics discussed and their relative importance. Overall, this study provides a comprehensive overview of the state of research in customer relationship management and can serve as a valuable resource for researchers and practitioners alike.

Recommendation and Managerial Implication

To enhance the impact and relevance of research in the field of customer relationship management, it is recommended to increase research collaboration among institutions and countries.

Increase Research Collaboration: With the increasing number of publications in the field of customer relationship management, it is essential to encourage collaboration among researchers from different institutions and countries. This can help to develop a more comprehensive understanding of the subject, identify research gaps, and produce innovative solutions.

Focus on Emerging Trends: As the number of publications is declining in recent years, it is crucial to identify emerging trends and topics in customer relationship management to keep the research relevant and up-to-date. Managers and researchers can collaborate to identify these trends and develop research projects that align with the current needs of the industry.

Encourage Diversity: The analysis of the most prolific authors and institutions reveals a lack of diversity in terms of gender, race, and geographic location. To promote diversity in the field, managers and researchers can encourage the participation of underrepresented groups in research projects and conferences, provide training and support for these groups, and create a more inclusive environment for collaboration and innovation.

Improve Research Quality: With the majority of publications being articles and conference papers, it is crucial to ensure the quality of research by following ethical standards, conducting rigorous research design and data analysis, and providing clear and transparent reporting. Managers and researchers can collaborate to develop best practices and guidelines for research quality assurance, and encourage the adoption of these practices in their institutions and research projects.

Increase Funding Support: The analysis of the Top 10 sponsorship shows that funding support plays a critical role in the production and dissemination of research in the field of customer relationship management. To promote further research and innovation, managers and researchers can explore different funding opportunities, such as government grants, industry partnerships, and philanthropic donations, and use these resources to support research projects and collaborations.

References

1. Khan, R.U., Salamzadeh, Y., Iqbal, Q., Yang, S.: The impact of customer relationship management and company reputation on customer loyalty: the mediating role of customer satisfaction. J. Relatsh. Mark. **21**, 1–26 (2022)
2. Haryandika, D., Santra, I.K.: The effect of customer relationship management on customer satisfaction and customer loyalty. Indones. J. Bus. Entrep. (2021). https://doi.org/10.17358/ijbe.7.2.139
3. Yoong, L., Lian, S.: Customer engagement in social media and purchase intentions in the hotel industry. Int. J. Acad. Res. Bus. Soc. Sci. **9** (2019). https://doi.org/10.6007/IJARBSS/v9-i1/5363
4. Kushwaha, B., Singh, R.K., Tyagi, V., Singh, V.: Ethical relationship marketing in the domain of customer relationship marketing. Test Eng. Manag. **83**, 16573–16584 (2020)
5. Hassan, H., Bin-Nashwan, S.: Impact of Customer Relationship Management (CRM) on customer satisfaction and loyalty: a systematic review. Res. J. Bus. Manag. **6**, 86–107 (2017)

6. Dogan-Sudas, H., Kara, A., Cabuk, S., Kaya, K.: Social media customer relationship management and business performance: empirical evidence from an emerging market. Stud. Bus. Econ. **17**, 90–107 (2022). https://doi.org/10.2478/sbe-2022-0027
7. Ryals, L., Payne, A.: Customer relationship management in financial services: towards information-enabled relationship marketing. J. Strateg. Mark. **9**, 3–27 (2001). https://doi.org/10.1080/713775725
8. Power, D.: Decision Support Systems: Concepts and Resources for Managers (2002)
9. Jafari, T., Zarei, A., Azar, A., Moghaddam, A.: The impact of business intelligence on supply chain performance with emphasis on integration and agility–a mixed research approach. Int. J. Prod. Perf. Manag. ahead-of-print (2021). https://doi.org/10.1108/IJPPM-09-2021-0511
10. Turban, E.: Decision Support and Business Intelligence Systems. Pearson Educ, India (2011)
11. Filip, F.G., Zamfirescu, C.-B., Ciurea, C.: Decision Support Systems. In: Filip, F.G., Zamfirescu, C.-B., Ciurea, C. (eds.) Computer-Supported Collaborative Decision-Making, pp. 31–69. Springer, Cham (2017)
12. Alavi, M., Leidner, D.: Review: knowledge management and knowledge management systems: conceptual foundations and research issues. MIS Q. **1**, 107 (2001). https://doi.org/10.2307/3250961
13. Gebert, H., Geib, M., Kolbe, L., Brenner, W.: Knowledge-enabled customer relationship management: integrating customer relationship management and knowledge management concepts. J. Knowl. Manag. **7**, 107–123 (2003). https://doi.org/10.1108/13673270310505421
14. Demirkan, H., Delen, D.: Leveraging the capabilities of service-oriented decision support systems: putting analytics and big data in cloud. Decis. Support Syst. **55**, 412–421 (2013). https://doi.org/10.1016/j.dss.2012.05.048
15. Hsieh, C.Y., Su, C.C., Shao, S.C., et al.: Taiwan's national health insurance research database: past and future. Clin. Epidemiol. **11**, 349–358 (2019). https://doi.org/10.2147/CLEP.S196293
16. Pio, L., Cavaliere, L.P.L., Khan, R., et al.: The impact of customer relationship management on customer satisfaction and retention: the mediation of service quality. Türk Fiz ve Rehabil Dergisi/Turkish J Physiother Rehabil **32**, 22107–22121 (2021)

Deep Learning Techniques for Wireless Networks and Data Analysis

Wireless Sensor Network-Based Wireless Safety System Using Underground Mine Workers

S. Gopalakrishnan(✉), K. Rani Swetha, S. Manisha, and P. Sreenath Reddy

Department of Electronics and Communication Engineering, Veltech Rangaraja Dr. Sagunthala R&D Institute of Science and Technology, Chennai, Tamil Nadu 600062, India
drsgk85@gmail.com

Abstract. The world's most dangerous location to work is in a mine, since there are frequently explosions there that result in thousands of fatalities. And according to a current study, these mining natural disasters have claimed the lives of, on average, around 200 individuals. In the proposed system, an IOT-based underground mine safety monitoring and alert system A concept is proposed to establish security and detect accidents in an underground mine, this is among the most significant IOT undertakings. The two main sections of the work are the receiver and transmitter sides. The transmitter sides are temperature, smoke, and gas sensors. The transmitter section is also based on an enabled device component. An IOT screen is included in the receiver section. All sensor data is delivered to a wired IOT server in every section through a Node MCU controller. The DHT11 humidity sensor emits an alert when the threshold limit is exceeded. Miners have one of the lowest-risk occupations in the underground application. In certain nations, underground miners are not guaranteed social protection or safety, and they may be left on their own in the event of an injury. The receiver and transmitter modules make up the project's two main components. Humidity sensors and smoke sensors are incorporated into the transmitter module. A Wi-Fi module is also included in the transmitter module. The receiver module includes an LCD screen. Through a Wi-Fi module, all sensor data is sent every two minutes to a distant IOT server. An indication that occurs when the threshold limit rises is also a part of the circuit. One of the lowest-risk professions in the world is mining. In certain nations, underground miners are not guaranteed social protection or safety, and they may be left on their own in the event of an injury.

Keywords: Temperature Sensor (DHT11) · Humidity Sensor · GAS Sensor · Node MCU ESP8266 · LCD Display

1 Introduction

Underground mining operations show to be an adventure in terms of worker safety and health. These risks exist as a result of the various methods used to extract important minerals. As the mine's depth grows, so does the risk. These safety precautions are particularly important in the coal business. Hence, whether mining coal or other minerals, worker safety should always come first. Underground coal mining is considerably riskier than open pit mining due to ventilation concerns and the chance of a collapse. All types of mining, however, include safety concerns owing to the use of heavy gear and excavation processes.

Modern mines generally implement various safety protocols, worker education and training, and health and safety requirements, which fundamentally change and improve safety standards in both opencast and underground mining. Coal is one of the most important better and raw resources for several industries. It is used to produce power in addition to extracting several chemicals and minerals from byproducts. But so far, extracting coal from the mine is a complex and hazardous process. Several coal mining accidents have occurred globally, causing lives as well as monetary damages. Recent smart technologies can reduce the dangers and hazards. Figure 1 develops a prototype system in this work that has a broad coverage area and enables us to access the situation from every server and continuously track underground data.

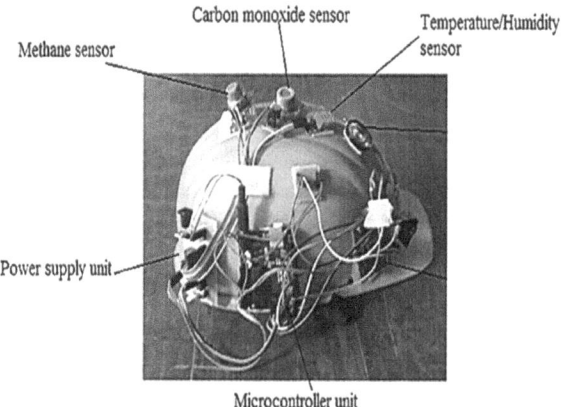

Fig. 1. Model prototype for Underground mine safety for workers

1.1 Objective

IoT is implemented in the mining industry to maximize effectiveness, strengthen safety protocols, and spread awareness. Monitoring the safety of miners if an increase in temperature, pressure, or gas occurs in a mining accident to maintain surveillance of the conditions in the mines and alert the workers in case of an emergency. The miners are alerted when hazardous gases are detected and when an accessory is removed.

2 Previous Research Work

Internet of Things (IoT) is a group of gadgets (things) that are connected to the internet. Smart cities as a whole encompass clever housing, smart transportation, and smart housing, to name a few. The internet of things (IoT) is a game-changing breakthrough that enables all sensory data to be stored in the cloud and accessed fast. The observed data is collected by the Arduino UNO and stored in the cloud. If the mining parameters exceed the threshold level, an alert message is conveyed to the control room. The buzzer will sound to alert the miners if even the slightest flame is detected fuel sensor, MQ11 (flame sensor), and THD11 (temperature and humidity sensor) [1].

Major controller unit, managing all physical components using a Python-based software interface. The raspberry pi device is connected to the MQ2 gas sensor, DHT11 sensor, LDR sensor, Ultrasonic sensor and motor driver unit. The Raspberry Pi starts executing the code as fast as the system is switched "ON," testing to see whether any hardware components are connected to the system before every component begins to perform its function [2].

The ensemble learning prediction model, which combines many algorithms, is employed. The data is received by the application via an IoT setup that takes into account several characteristics such as temperature, humidity, fire, and gas concentrations. The Arduino Uno, which is linked to several sensors, is used to gather data highlights such as temperature, humidity, gas, and fire. The samples are collected in the CSV record format using the PLX-DAQ programmer [3].

The technology user's offer is specifically intended for coal mines, where mishaps are prevalent due to pit gas concentration and rapid ignition. It allows for continuous monitoring of crucial environmental elements as well as surveillance of humans. To monitor methane concentrations and keep surveillance on the employees in the major coal mine galleries, the communication system described here intends to enable data collecting from below and transport them, in real-time, to the mine surface [4].

A digital gadget that uses ZigBee communication technology, an accelerometer, and a dust sensor to communicate collected data. Taeyoung EMC deployed 29 vibration modules, 14 dust modules, and 2 coordinator modules at its Samdo Mine in Samcheok, Republic of Korea. We're able to detect variations in vibration and dust before and after blasting [5].

A system that can measure the temperature, air quality, gas level, and pulse rate of a worker. Sensors are used to measure the amount of gas and humidity in the air. The carbon monoxide demarcation level, as well as a humidity sensor that detects the moisture in the surrounding air, detects gas. In this suggested system, IoT is employed for data transmission from below ground to the management system as well as data receiving [6].

Fundamental task of a VoIP-based communication system is to compress voice packets before transferring them over the TCP/IP or UDP protocol to the receiver. The hardware and software components of the proposed communication method have been divided into two parts. In the hardware section, the Raspberry Pi's microphone, keyboard, battery, and speaker modules are configured such that they are strong, easily portable, and connected securely [7].

Gas sensor modules, temperature sensors, water level sensors, and relays are used to fix the coal mine safety systems. Attach the controller to every sensor. To proceed, should create a Thing Speak account. Also, have monitoring and controlling technologies in this system. All of the information collected by different sensors is analyzed by the monitoring system. In this project, a low-cost wireless mine supervision system with early-warning intelligence is proposed. IOT can be utilized to monitor worker status [8].

System for continuous monitoring that really can monitor issues like the presence of harmful gases (methane, liquid gas, and Carbon monoxide), humidity and temperature, enabling detection of helmet utilization, Scale on the skull has been damaged. In general, this area is in charge of improving safety (obviously), as well as data evaluation, processing, and transfer. Three elements make up the smart safety helmet: assessments, data transfer, and alarm [9].

A conceptual framework for IoT implementation to improve the Smart-SAGES underground support system. To clearly define security needs, a detailed overview of the taxonomy of security challenges in the UMC IoT combination has been provided. Due to the diverse equipment and distributed network, the block chain-based system appears to be a potential solution for the mining sector. It may aid in preventing the penetration and disruption of cyberattacks [10].

Adequate and effective communication under the dynamic and unsafe conditions found in underground coal mines is a very difficult task. When it comes to initial investment expenses, PLC-based communication is a particularly appealing solution. To achieve this, it is necessary to carry out in-depth PLC experiments to develop a straightforward and approachable voice communication model for underground coal mine environments. Also, it is essential to incorporate PLC technology into the power line that is already set aside for emergency lighting systems [11].

Smart safety monitoring system is proposed that can position precise personal tracking in industrial fields like mines, underground operations, tunnel engineering, and underground engineering. The system is based on an improved DV-Hop localization algorithm for randomly deployed wireless sensor networks (WSNs). The original DV-Hop has been significantly altered by the selection of anchor nodes and the calculation of the average per-hop distance between anchor nodes [12].

A mine environment perception system that uses in coal mines is researched and developed. The sensing layer adopts a wireless sensor network based on Wi-Fi technology, utilizes the miner safety helmet as the carrier integrates a function module with an STM32F107 controller as the core, and becomes a mobile node of the underground wireless sensing network to monitor personnel positioning and the information relevant to the miner's workplace environment [13].

WSN-based safety monitoring in coal mines suggests a simple AKA scheme. By using the AVISPA tool, simulated the technique. Both formal and informal security analyses were conducted using the Random Oracle Model (ROM). These analyses demonstrate that the recommended strategy is safe and impervious to several well-known techniques. Regarding security characteristics and computational cost, compared the proposed method to other related schemes. In comparison with other systems, ours is more secure and has a similar computational cost [14].

Ground station is crucial in ensuring the quick rescue of miners. It is possible to use some of the technologies with modifications according to all the safety protocols made and the advancement of technology. To successfully monitor the UG mine environment, further research in wireless sensor networks and applications is surveyed in this study [15].

2.1 Problem Statement

When it comes to the health and safety of the workers, mining operations are a dangerous business. These hazards arise from different methods used to extract important minerals. Thus, in any kind of mining, worker and resident safety should always be a primary concern.

The difficulty is to find out how to digitalize these safety protocols, enabling data collection on the field, automating report generation, and also providing analytics that can let mine and central management evaluate, so in the proposed method developed as discussed below.

3 Materials and Method

In this work, Arduino is employed to obtain all output data from the sensors as mentioned an humidity sensor is a device specially designed to measure how hot or cold is it. An accurate IC temperature sensor with a yield that matches the gas like smoke. The DHT11 can calculate the temperature more precisely than a thermistor can. A gas finder is a device that searches for gases in a specific area, generally as part of a safety plan. This kind of device may communicate with a control framework and is utilised to find gas spills and other outflows. a thermistor and a humidity-detecting component work together to measure temperature. A capacitor works as the sensing component of a capacitive sensor. The change in the dielectric material's electric generates is monitored to determine the relative humidity values and over block diagram shown in Fig. 2.

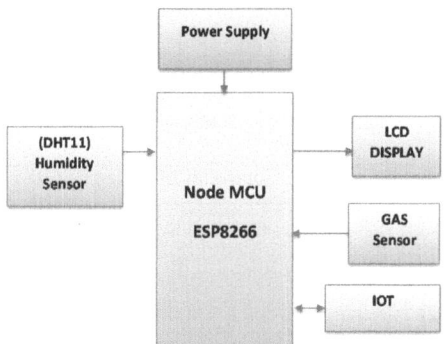

Fig. 2. Proposed block diagram

This Internet of Things-based method for maintaining the safety of underground coal mines is described, and the desired systems are spread out around the mine. Temperature, smoke, and LDR sensors are a few examples of both environmental and physiological aspects that the sensors detect. These data are transmitted to the controller so that the information may be shown on the LCD. The controller is used to control IOT to transmit information to the control system when sensors exceed the threshold level, and the control system alerts the mining work area whenever this happens. This system concentrates on analyzing the factors that can prolong life and transmits this data to the IOT server for immediate support.

3.1 Node MCU ESP8266

Figure 3 shows the ESP8266 is a strong and autonomous Wi-Fi network solution that can carry code applications and disable all Wi-Fi networking functions. After the device is installed and the appliance processor's only application, the non-volatile storage will be triggered immediately from an external Move. Cache memory incorporated with the system can boost performance and reduce memory requirements. Another circumstance is that once wireless internet access assumes the function of a Wi-Fi adaptor, it is simple to include it into any controller-based design using the hardware AHB bridge interface or the SPI / SDIO interface.

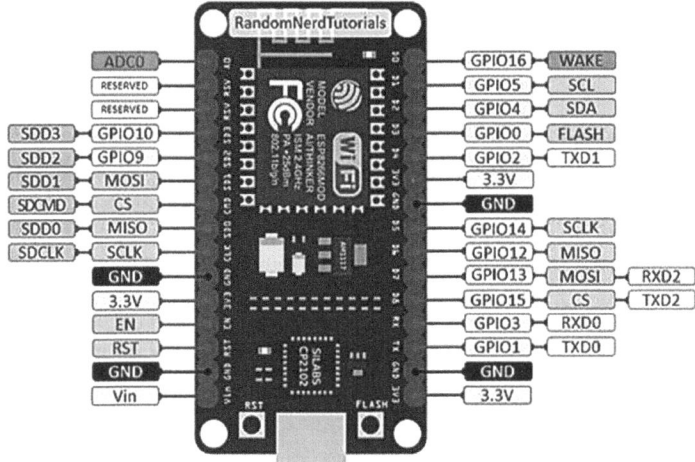

Fig. 3. Controller Module

3.2 Gas Sensor (MQ2)

For detecting gas spills, the Grove - Gas Sensor (MQ2) module is helpful (in the home and industry). H2, LPG, CH4, CO, alcohol, smoke, and propane will all be detected. Backed up its quick reactions. As soon as possible, measurements need to be taken. Figure 4 shows the potentiometer simultaneously adjusts the sensitivity.

1. Air quality monitoring is used to detect gases in the air such as butane, LPG, and methane.
2. A broad detection range.
3. Stability, durability, and high accuracy.

Fig. 4. Gas sensor

3.3 Temperature Sensor

Because of its low output impedance, a linear output, and exact intrinsic calibration, the LM35 is particularly easy to connect with reading or control circuitry. It is compatible with both positive and negative power supplies, as well as a single power supply. The output voltage of the LM35 series precision integrated-circuit temperature sensors is exactly proportional to the temperature in Celsius (Centigrade). The LM35 has an advantage over linear temperature sensors calibrated in Kelvin since it does not need the user to remove a substantially constant voltage from its output to obtain suitable Centigrade conversion. Figure 5 shows the LM35 does not require external calibration or trimming and has typical accuracies of 14 °C at room temperature and 34 °C throughout the whole temperature range of 55 to + 150 °C. Low cost is ensured by cutting and calibrating at the wafer level.

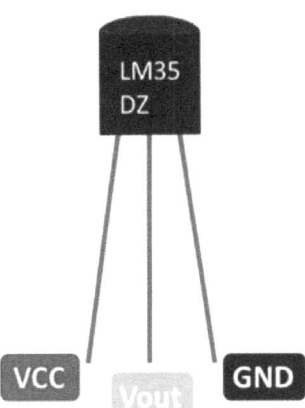

Fig. 5. Example module of LM35 Temperature sensor

3.4 Humidity Sensor (DHT11)

Moisture in the air is really what causes the environment to become humid. In addition to affecting various industrial production processes, the amount of water vapor in the air will also affect human comfort. Many physical, chemical, and biological processes are impacted by the presence of mist. Activity related to humidity in the industrial sector is significant since it will have an impact on both the staff's health and safety as well as the commercial cost of the better. As a result, humidity detection is critical, particularly in industrial processes and human comfort management systems. Figure 6 shows the commonly used DHT11 temperature and humidity sensor equipped with an 8-bit CPU that produces temperature and humidity values as serial data, as well as a specialized sensor. NTC for temperature measurement.

Fig. 6. Example module DHT11 humidity sensor

3.5 LCD Display

One of the technological developments that are frequently used as displays in screens, phones, work-stations, and other devices is the liquid crystal display (Fig. 7). It combines the powerful and fluid characteristics of a problem. LCD uses fluid gem to display a clear image. Presentations may be much more slim on LCDs than on CRTs.

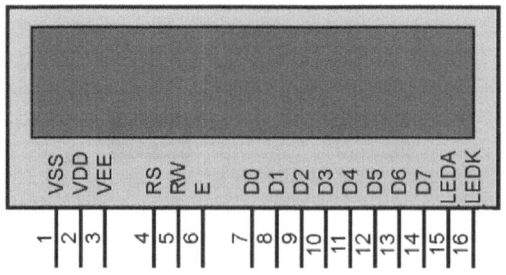

Fig. 7. Module of 16x2 LCD

Instead of emitting light, they use backdrop illumination to create shady or monochrome visuals. The pixels on an LCD are illuminated by a backdrop lighting system. It has red, blue, and green subpixels that may be activated or deactivated using pixels. When every single subpixel is turned off, the presentation seems black, and when every subpixel is switched on, it appears white. Only 50% of the light is transmitted through the layers of the fluid precious stone showcase's background lighting, which spellbinds the light. The powerful and fluid elements of the presentation can be bent by applying voltages. This assists in turning on and off the showcase's lights.

4 Result and Discussion

Users can implement the full project based on mine workers for safety, security devices using an improved technique that depends on IOT. To detect whether or not rocks are falling, the tilt sensor is used. DHT11 and gas sensors measure the temperature and gas; if any gas value detects gas leakage, which could be dangerous, and the temperature rises, obtain significant accuracy numbers from this and have quicker access to the data on the server. Then present the results on the parameter screen for monitoring. The buzzer will turn on after the alarm threshold has been passed by the sensing element values (Fig. 8).

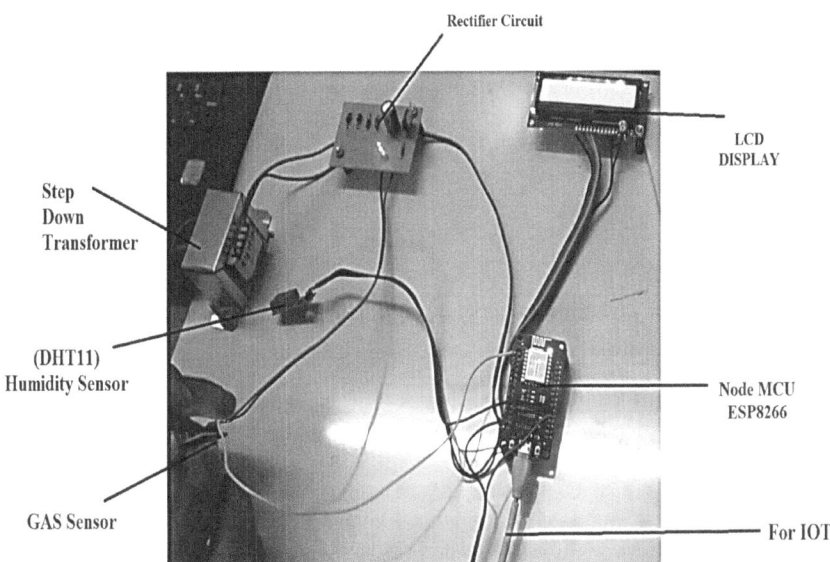

Fig. 8. Proposed Hardware Module based Underground mine safety

A step-down transformer is a component of electrical devices that reduces the voltage of an AC power source. It is composed of an iron core, primary winding, and secondary winding. An oscillating magnetic field is generated in the iron core when an AC voltage is delivered to the primary winding. A voltage is then generated in the secondary winding by this magnetic field, however, it is less intense than in the main winding.

5 Conclusion and Scope for Future Work

The proposed system was created and tested to demonstrate the feasibility of its prototype. A wireless sensor network-based system for monitoring coal mine safety, as well as its software and hardware design, are extensively addressed in the current study. This prototype monitors three parameters (humidity, and gas) in a coal mine and automatically displays an intimation on the LCD when the environment parameters are abnormal and exceed the threshold, which helps enhance monitoring safety and decrease accidents. As a Concluded output of the test experiment's positive results are achieved used in large industrial or deep mining environments.

In Future monitoring of all potential safety hazards, including gases, dust, vibrations, fire, etc., might be done with the help of additional sensors. Moreover, Zigbee may be used to monitor mining operations for things like subsidence and water leaks. Cable communication may be a problem since other crucial data may be transmitted through this method. Because of the system's simplicity of use, the control may be managed from the surface itself. In the mining environment, there is a significant risk that can do a great deal of harm. The mine safety system's re-suppression subsystem is another that may be included. In this, a smoke sensor and a regulatory fire retardant system are going to be utilized.

References

1. Ansari, A.A., Gera, P., Mishra, B., Mishra, D.: A secure authentication framework for WSN-based safety monitoring in coal mines. Ind. Acad. Sci. (2020)
2. Chen, K., Wang, C., Chen, L., Niu, X., Zhang, Y., Wan, J.: Smart Safety Early Warning System Of Coal Mine Production Based on WSNS. Elsevier ltd. (2020)
3. Raul, R.N., Palit, S., Maity, T.: Challenges and Viability of the Use of PLC For Personal Communication in Underground Coal Mines. Springer (2020). ISSN: 0256-4602
4. Rout, R., Pramanik, J., Das, S.K., Samal, A.K.: Alleviation of Safety Measures in Underground Coal Mines Using Wireless Sensor Network: A Survey. Springer (2020)
5. Stoicuta, O., Riurean, S., Burian, S., Leba, M., Ionica, A.: Application of Optical Communication for an Enhanced Health and Safety System in Underground Mine. Mdpi (2023)
6. Harshitha, P., Dixith, H.S., Krithi, Rajeeva, N., Shashikala: Coal mine disaster prediction. International Journal of Engineering Research and Technology (IJERT), **9**(07) (2020). ISSN: 2278-0181
7. Lee, W.-H., Kim, H., Lee, C.-H., Kim, S.-M.: Development of Digital Device Using ZigBee for Environmental Monitoring in Underground Mines. Mdpi (2022)
8. Dey, P., et al.: Deep convolutional neural network based secure wireless voice communication for underground mines. J. Amb. Intell. Human. Comput. (2021)
9. Renuka, N., Saisree, P., Chandana, S., Salman, MD., Deepak, B.: Iot-based underground worker safety system. Int. J. Res. Appl. Sci. Eng. Technol. (IJRASET), **10**(11) (2022). ISSN: 2321-9653
10. Ansari, A.H., Shaikh, K., Kadu, P., Rishikesh, N.: IoT based coal mine safety monitoring and alerting system. Int. J. Sci. Res. Sci. Eng. Technol. **8**(3), 404–410 (2022). ISSN: 2395-1990
11. Manohara, K.M., Nayan Chandan, D.C., Pooja, S.V., Sonika, P., Ravikumar, K.I.: IoT based coal mine safety monitoring and alerting system. Int. J. Eng. Res. Technol. (IJERT), **10**(11) (2022). ISSN: 2278-0181

12. Herur, S., Leema, C., Gowri, M.G.: IoT-based coal mine safety monitoring and control automation. Int. J. Eng. Res. Technol. (IJERT) **10**(11) (2022). ISSN: 2278-0181
13. Henriques, V., Malekian, R.: Mine Safety System Using a Wireless Sensor Network, vol. 4. IEEE (2020)
14. Srivastava, S.K.: Real-time monitoring system for mine safety using wireless sensor network (multi-gas detector) (2020)
15. Ghadyani, D., et al.: Real-time monitoring and alarm system in underground coal mines using smart helmets (a case study: tabs coal mine). Res. Square (2021)
16. Singh, A., Kumar, D., Urgen, J., Otzel, H.: Iot based information and communication system for enhancing underground mines safety and productivity: genesis, taxonomy and open issues. Ad-hoc Netw. (2020)
17. Chen, W., Wang, X.: Coal mine safety intelligent monitoring based on wireless sensor network. IEEE Sens. J. **21**(22) (2021)
18. Porselvi, T., Sai Ganesh, C.S., Janaki, B., Priyadarshini, K., Shajitha Begam, S.: IoT based coal mine safety and health monitoring system using lorawan. In: International Conference on Signal Processing and Communication (2021)
19. Lande, S., Chabukswar, P., Bhope, V.: An efficient implementation of wireless sensor network for performing rescue and safety operation in underground coal mines. In: International conference for engineering technology (2020)
20. Soomro, A.H., Jilani, M.T.: Application of IoT and artificial neural networks (ann) for monitoring of underground coal mines. In: International Conference on Information Science and Communication Technology (ICISCT), pp. 1–8 (2020)

A Meticulous Analysis on Energy Based Routing Protocol in Underwater Sensor Networks

B. Anandha Mathavan[(✉)], A. Shenbagharaman, B. Paramasivan, and B. Shunmugapriya

National Engineering College, Thoothukudi District, Kovilpatti, Tamilnadu, India
{2151009,sraman,bparamasivan,bsp}@nec.edu.in

Abstract. Underwater wireless sensor networks (UWSNs) are presently a hot exploration affair in academe and commerce, with numerous undersea functions including tide examination, seismic examination, coincidental examination, and ocean floor examination. However, UWSNs face a statistic of constraints and protests, along with immense sea intrusion and blast, long breeding delays, cramped radio band, dynamic grid topology, and sensor node battery spirit limitations. One solution to these dispute is the development of annexation protocols. A annexation protocol can transfer information from the expert apex to the target apex in the network in an efficient manner. This article examines undersea routing protocols for UWSNs. Extant underwater annexation covenants are restricted into three categories: spirit-established, evidence-established, and geological information-established covenants. In this article, we outline the proposed underwater routing covenants of spirit based protocol. The proposed protocol is meticulously explained, along with benefits, drawbacks as well as mathematical modeling. Meanwhile, a meticulous analysis of the effectiveness of several underwater routing protocols is being done. In extension, we confer underwater annexation methods research obstacles and potential possibilities, which can aid future exploration.

Keywords: Sprit-established · evidence-established · geological–information established · routing

1 Introduction

Tides contain approximately 96% of all clay water, which is critical for humanity's continuity because it provides innate assets, coastal defense, and other perk [1]. Because of overdevelopment of area, institute and analysts have shifted their attention to the tides. Ocean monitoring and research are challenging undertakings, despite the ocean's abundance of resources. The underwater sensor apexes are placed in a certain territory of the ocean, and the nodes' capacity to self-organize is leveraged to transmit data, creating the UWSN [5]. Communication is the best critical network automation. Underwater wireless communication is the communication of information in a water situation using wireless technology [2]. Acoustic, Electro-magnetic (EM), and optical waves are examples of carriers. Many wireless communication methods that may be employed in a terrestrial context cannot be used effectively in aquatic conditions because of the unique flavor of the underwater habitat, such as pressure and temperature.

Electromagnetic waves can be effectively transported over an area. Seawater, on the other hand, significantly attenuates radio signals. Ultra-low frequency electromagnetic waves with a frequency range of 30 to 300 Hz are capable of penetrating extra than 100 meters of saltwater. It does, however, necessitate a lengthy collecting antenna, which a short sensor apex cannot provide. Therefore, walkman waves can only accomplish immense-momentum communication across a certain number of distances. Since there is a lot of bandwidth available, visual communication can deliver immense data rates (up to Gbps) inside a few tens of meters. The underwater optical wireless communication (UOWC) technology from BlueComm and Ambalux allows for data transfer across a distance of around 100 meters. According to the study, LED-based visible light communication is feasible in clear water across a 500-metre link distance. A high-capacity wireless communication link can be created via visual communication, which has a steady information-carrying capability. The light signal is, however, heavily engaged and diffused, thus the UC's communication quality is highly tied to the purity of the water. The most often employed technology is acoustic communication. Acoustic flakes are mechanical flakes with a large transmission range that can be used for warmth-stable buried-water conversation (up to tens of kilometers or a few hundred meters) [3]. Undersea conversation, the most advanced undersea conversation technology, is the sole way to enable underwater cellular long-orbit conversation amid undersea objects. However, as lofty- prevalence acoustic flakes travel through seawater, they will experience significant attenuation, leaving just a small amount of bandwidth available. Compared to the speed of RF waves, the communication momentum of acoustic flakes is roughly five orders of significance slower. Communications underwater will be severely delayed as a result. Acoustic wave transmission is best supported in seawater because its attenuation is substantially smaller than that of RF and optical waves [4].

2 Related Work

Related work to this publication addresses various routing strategies created specifically for underwater wireless sensor networks. Spirit-established protocols, evidence-established protocols, and geological information-established protocols are the three groups into which the aquatic routing protocols are divided. Established on node spirit, data transfer, and geographic placement criteria, it is evaluated. Since the environment for spirit-based protocols is particularly complicated, the undersea sensor nodes' batteries cannot be constantly changed. Energy constraint is thus the key component of the aquatic routing systems. Established on the energy available at the sensor apex (the node's current residual point) or the spirit appeal on the transmission grid, the spirit-efficient annexation protocols choose the best suitable direction from the origin apex to the target apex for the information to be upheld (Fig. 1).

Protocols for Energy Management: Choosing the apex with the minimum cost and being aware of the later-hop apex on the path to the target node are both characteristics of an aware sensor node. Routing protocols established on spirit perception, which also find the optimum transmission route, can effectively address the problem of the spirit

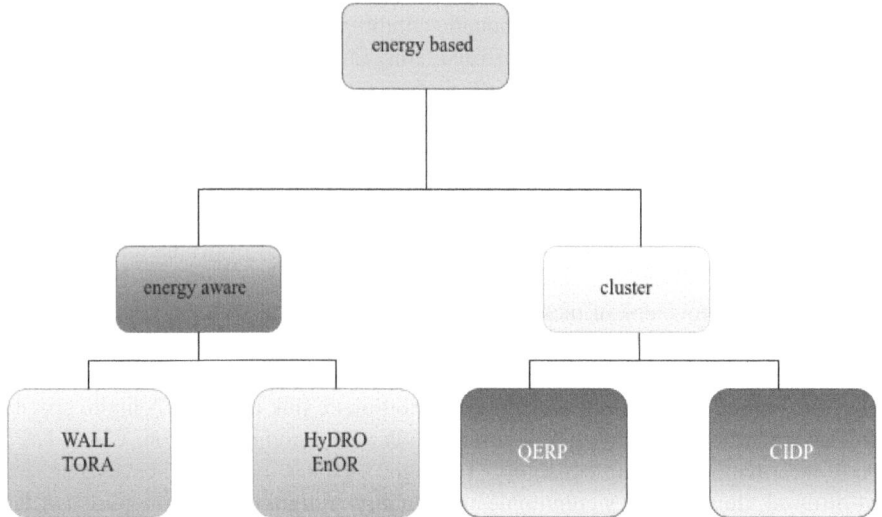

Fig. 1. Various energy based routing protocol in UWSN

consumption of the sensor apexes. The peeved-stack architecture of the subsea routing protocol can improve the lone-hop act.

2.1 Energy Aware Routing Protocols

WALL

The provided set of connections is used for uni-cast routing in a wireless acoustic line link (WALL) routing strategy. Both network longevity and quality of service are guaranteed. The cross-stack model will be used by each node, giving the grid layer a way to determine the throughput of a conversation channel. The grid layer will configure spirit-efficient annexation with the continued end-to-end container lag using information on throughput and spirit utilization [6].

The information-link bed and the grid bed above cross the environmental layer in Zhiqiang Liu's unique cross-stack design. With the help of our cross-stack, we can analyze a set of links that either achieves or exceeds a given throughput benchmark. It will go up the various components of our peeved-stack model, focus several suppositions, and list the additional QoS metrics, including throughput and spirit, that our routing method uses. The current cross-stack model is intended to design the environmental and information-link layers simultaneously for the best lone-hop grid throughput. The CSMA/CA protocol, also known as slotted Multiple carriers perceive access, is used in this study.

TORA

The TORA protocol is mainly worn to diminish end-to-end lag, handle the dispute of empty apexes, minimize horizontal communication, boost network throughput, and improve energy efficiency. Its three operating steps are apex location, aspirant forwarder election, and information communication. Utilizing the TOA (Time of Arrival) and dimension, network nodes are found. The excellent uphold node that is closest to the target is elected using the locale suits and residual spirit of the apexes. TORA [7], Ziaur Rahman's receiver-based opportunistic and geographical routing system, is a hierarchical localization technique. To locate the optimum forwarding node that is closest to the target, TORA uses apex localization techniques like trilateration and TOA, as well as residual spirit from the apexes.

The three steps of this scheme's functioning are localization of nodes, candidate forwarder selection, and data transfer.

Nodes localization
Surface sonobuoys and regular sensing nodes are the two types of nodes that are utilized in this proposed strategy. The sonobuoys are outfitted with typical GPS units and are designed to receive their precise location from a satellite. A DCA (Double Communication Apex) or a LCA can be used like a regular node (Lone Communication Apex). During the initial stage, normal apexes that are inside the sonobuoys' transmission range are designated as LCAs. DCAs can connect with LCAs to estimate their position in the network even though they are not meant to be inside the sonobuoys' broadcast range. These nodes use sonobuoys to determine their location and help the other nodes localize. The multi-sink architecture of the proposed approach calls for the deployment of three or more sink apexes at the water's facial. Every node has a pressure sensor, and by using that sensor, ordinary nodes can determine their depth. The 3-dimensional localization problem can be reduced to a 2-dimensional problem once you know the depth of a node, and a range-based localization strategy can be used to forecast the coordinates of a node's location. This method can trilaterate after a range-based estimate to achieve a normal node position. Each node needs to receive position data from at least three reference nodes in order to perform trilateration.

Following that, the node localization step begins with a sink hello message. Every so often, sinks positioned at the water's surface convey greetings. This message comprises specific parameters that aid common apexes in locale, such as "Bed ID," "APEX ID," "APEX Status," "Locale suits," "Optimism Value," "Level ID," and "Proximity to Sink." The type of a packet describes whether it is a control packet or a data packet and if it was initiated by Sonobuoys or ordinary nodes. Apex ID is a specific ID to any apex, whereas Apex Dignity identifies the breed of apex, such as sonobuoy, LCA, DCA, or non-localized apex. Locale suits show the precise locale of an apex. A normalized version of location error known as the confidence value will be described in more depth below. The sink hello packet will have zero values for Proximity to Sink, Bed ID, and Confidence Value; LCA will use these values to continue the localization process. The orbit between the expert and sink nodes can be estimated if a container is sent through an acoustic salient at moment T1 and it arrives at the sink by moment T2.

$$Dis.2sink = V(T2 - T1) \tag{1}$$

Where V is the acoustic salient propagation momentum.

The orbit between the sender and receiver is calculated using Eq. (1). A node utilizes trilateration to determine its coordinates after it is aware of its length to three reference nodes. To get a node's specific locale suits, the three allusion apexes should be selected so that the intersection of the three planes created by the overlap of the three spheres occurs at each node.

$$(x - k_1)^2 + (y - l_1)^2 + (z - m_1)^2 = r_1^2 \tag{2}$$

$$(x - k_2)^2 + (y - l_2)^2 + (z - m_2)^2 = r_2^2 \tag{3}$$

$$(x - k_3)^2 + (y - l_3)^2 + (z - m_3)^2 = r_3^2 \tag{4}$$

Where (k_1, l_1, m_1), (k_2, l_2, m_2) and (k_3, l_3, m_3) are suits of three allusion apexes, whereas (x, y, z) are suits of non-sectarian apexes and r_1, r_2, r_3 are orbits between non-sectarian apexes to three allusion apexes. This approach makes the assumption that a pressure sensor can be used to determine the intensity of a sensor apex. With the use of TOA, the distance to reference nodes has already been computed, and a pressure sensor may be used to determine a node's c-axis. As a result, it has three equations and two unknown variables, a and b. Now that it has solved these equations, it can determine the non-localized LCA's coordinates. LCAs' confidence value is set to one once they are localized close to sonobuoys. Once a non-localized node knows the coordinates of its position and how far it is from three LCA. It localizes itself via TOA and trilateration, and if its confidence value rises beyond a certain threshold, it can turn into a LCA. It will be labeled as DCA if not.

This scheme location mistake is denoted by the symbol δ, and it may be determined using the formula,

$$\delta = \sum_{k=1}^{3} (x - k_1)^2 + (y - l_1)^2 + (z - m_1)^2 - l_k^2 \tag{5}$$

Where (k, l, m) are the suites of reference apexes and (x, y, z) are the suites of a non-localized apex. With the use of Eq. (1), l_k is the orbit between allusion apexes and a non-localized apex. The author uses the assigned version of locale error, known as confidence value (ξ), to specify the node's state. The author determined a confidence value by

$$c = \left\{ 1 - \frac{\delta}{\sum_{k=1}^{3}(x - k_1)^2 + (y - l_1)^2 + (z - m_1)^2} \right\} \tag{6}$$

It has established the threshold value. Unknown nodes (ξ) can become reference nodes if their status is set to LCA and they are greater than φ; otherwise, they would become DCAs (Fig. 2).

Candidate forwarder selection

In this approach, the next forwarder is dynamically chosen based on particular matrices. Each receiving node is a potential candidate for forwarding a data packet once it has been broadcast by a sender. As a candidate matrix, apexes residual spirit and its orbit from the sink are employed. Therefore, the likelihood of the next forwarder being an

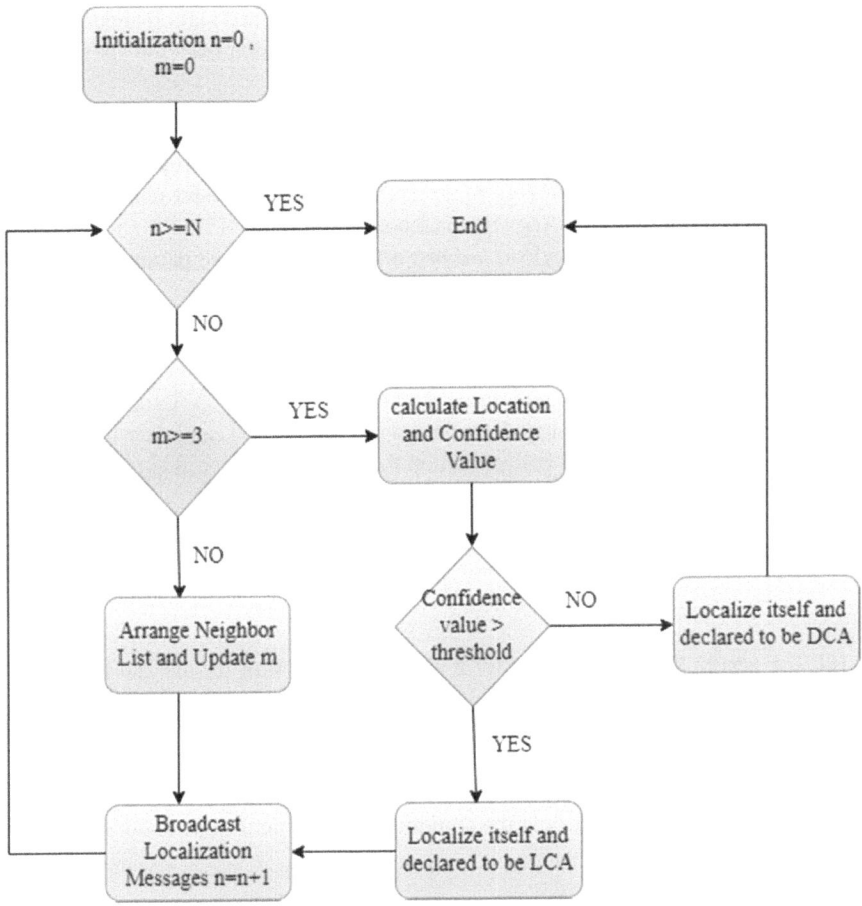

Fig. 2. Flow diagram of node localization

apex with larger residual spirit and nearby to the sink is very high. A node has a shorter holding period, as

$$H_{time} = \left(1 - \frac{E_r}{E}\right) * \left[1 - \left(\frac{D_{S,F} - D_{S,S}}{R}\right)\right] * T_{delay} \tag{7}$$

Where,
  E – initial spirit
  E_r - residual spirit.
  $D_{S,F}$ - the orbit amid a expert and forwarding node.
  $D_{S,S}$ - the orbit amid the expert and the sink node.
  R - superlative communication range
  Delay - predefined superlative delay

Delay is configured so that before transmitting their own information about the same packet, all apexes in the uphold set can hear the communication from the apex with the highest priority.

Statistics Communication TORA is built to prevent horizontal communication, minimize end-to-end latency, get rid of vacant apexes, and increase throughput and spirit efficiency. A crucial part of achieving these goals is the data transfer phase. Container Type, Sequence count, Apex Id, Apex Status, Locale Suites, Level ID, Proximity to Sink, Hop Count, and Information Payload are just a few of the unique parameters that are included in each data packet.

This method assigns the task of directing the packet along the best routing path to its destination to a candidate forwarder, who is chosen instantly. The broadcast characteristic of the medium is utilized by the opportunistic routing system TORA. Usually, there will be a number of potential forwarders. Specialized matrices called candidate sets are used to order the list. Once an expert collects 2-hop Ack for a container, it is treated to have been well endowed.

$$HF - T = 2 * \left[\left(\frac{R}{V} \right) + T_{delay} \right] \tag{8}$$

Where,

HT − F serves the equity period for a uphold apex to wait for the 2nd Hop apex's Ack.

R - superlative communication range for a apex

V - acoustic salient propagation momentum

T_{delay}- is set so that Before transmitting their own information for the same packet, all apexes in the upholding set can hear the communication from the apex with the greatest priority.

$$HT - SN = 2 * HT - F \tag{9}$$

Where,

HT − SN shows Senders believe that a packet is lost and start resending the same box when their holding time, which is set to twice the forwarder's equity time, expires without receiving an Ack (Fig. 3).

EnOR

EnOR - Energy-aware Opportunistic Routing is a brand-new, lightweight opportunistic routing technology that Rodolfo W. L. Coutinho suggests undersea sensor networks [8]. The goal of this covenant is to increase the grid lifetime, which is here defined as the percentage of live nodes over time. When beginning the aspirant set and prioritization algorithms, EnOR takes into account each neighbor's remaining energy, the progress (packet advancement), and the link reliability to its neighbors. Aspirant set selection and candidates' transmission prioritizing are EnOR's two key tactics.

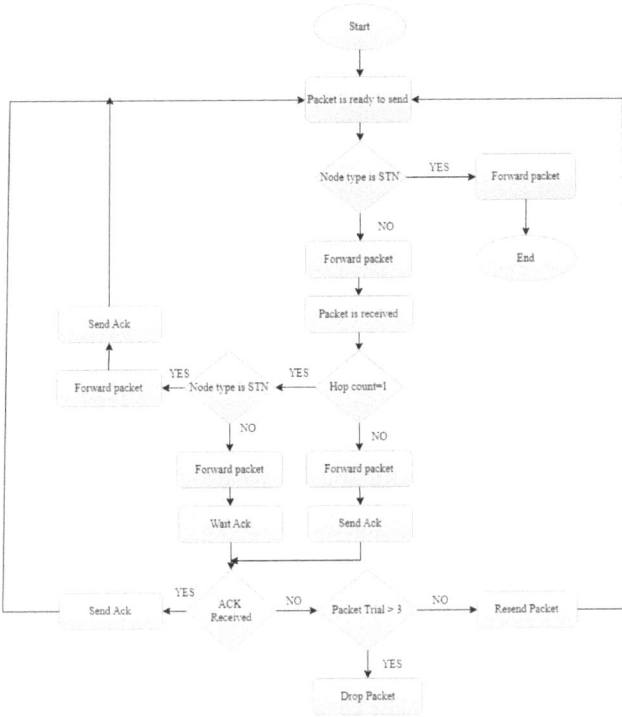

Fig. 3. Flow diagram of data transmission

Aspirant Set election Approach

The later-hop forwarding aspirant set is chosen by EnOR taking into account the loyalty, halting spirit, and container growth of the nearby apexes.

Intermittent beaconing

Beacon packets are routinely released by each underwater sensor node. A apex can calculate the pair wise orbit to the sender using approaches such as the collected salient strength indicator or Time of Arrival (TOA).

Selection of the Aspirant nodes
Procedure:

- Construe i as a sensor apex with a information container that has to be delivered, and define N as the table that is later to it.
- To choose the best suitable apexes as forwarding aspirants, the apex will compute the fitness value of its nearby apexes.
- An adjacent node is only regarded as an aspirant apex if it sends the container in the direction of the exterior sonobuoys.
- The intensity of the current sender apexes i and j is worn to determine the packet improvement of a neighbor j as $PA_j = depth(i) - depth(j)$.
- The nodes that have been approved as candidates are arranged in order of fitness value.

- From the prospective aspirant apexes, the collection of aspirant apexes is then selected. It is crucial to consider how many nodes are in the aspirant set. Low link reliability could be caused by a narrow aspirant pool.
- Up until the aspirant set meets the necessary connection dependability γ, potential aspirant apexes are added.

$$F_j = PA_j \times p(d_j, m) \times \left(\frac{E^j_{rem}}{E^j_{init}} \right) \qquad (10)$$

Where,

$PA_j > 0$ is the container growth of the acquaintance j.

$p(d_j, m)$ - is the delivery likelihood of an m-byte information container against apex i to apex j.

E^j_{rem} - halting spirit of j.

E^j_{init} - antecedent value of the spirit of j.

The link reliability of a considered set as,

$$P_d = 1 - \prod_{j=1}^{|r|} \lceil 1 - (p(d_j, m) \rceil \qquad (11)$$

Transmission priority level of aspirant apexes

EnOR naturally addresses the issue of all candidates having the same communication preference level. EnOR calculates the aspirant apexes' fitness value while taking into account their residual spirit to adjust the aspirant apexes priority on a periodic basis. Therefore, even if a node is the best aspirant for packet advancement, the less power it has left over, the lower its fitness value. A node with a lot of energy left over will be given a high priority in this fashion.

Information container communication

Finally, the information container is broadcast by the active sender apex. It contains both the maximum distance D_{max} between itself and the forwarding aspirant nodes as well as their IDs. The nodes' ids are listed in the information container header in descending order of priority.

Aspirant's communication Allocation

It is one of the more arduous concerns in polite annexation. EnOR uses timer-established allocation to organize the aspirant communications. Utilizing timer-established tactics, each applicant is allotted a time window depending on their preferences. As a consequence, based on its priority, each aspirant node maintains the data packet for a set period of occasion (or slots). Before this time period cease, the aspirant apex will announce the information container if it doesn't hear the communication of the identical packet coming from a huge-priority apex.

Packet holding time

In order to avert lengthy preambles in EnOR, the container equity time, established on the priority of the apexes, should take into consideration the greatest orbit amid the sender and the aspirants. A aspirant apex retains a information container for a specific amount of time after receiving it T_h,

$$T_h(p) = \left\{\frac{R-D_{max}}{v}\right\}, \; if \; p = 1 \\ \left\{\frac{R+p*D_{max}}{v}\right\} if \; p > 1. \tag{12}$$

R - conversation range

D_{max} - superlative orbit amid the sender and the aspirants.

p - preference of the aspirant agnate to its environment in the descent of the packet.

v - robust breeding momentum in the water.

Packet Transmission Suppression

EnOR employs an alive elimination technique. When a high preference apex forwards the packet, all low priority apexes that hear the communication halt the packet's transmission.

2.2 Cluster Based Routing Protocols

QERP

The fundamental benefit of QoS-Aware Routing protocols is that they present independent modules for various QoS metrics that work in concert with one another. Greater dependability, less end-to-end lag, and a above container consignment scale are the goals of these strategies.

Muhammad Faheem presented a Quality-of-Service (QoS) Aware Evolutionary Routing Protocol (QERP) [9]. It also improves container consignment while lowering moderate end-to-end latency and total grid spirit utilization.

Generic Representation of Scheme

The IDs of the genes (CH nodes) that a annexation path passes through are represented by the nonnegative number sequences that make up this system, which is referred to as a chromosome (routing path). The dimension of any established annexation path is configurable and must not exceed the entire number of declared CH apexes in order to prevent information path loops in the grid. The route with the highest residual spirit, SNR, the fewest number of member apexes, and the closest proximity to the sink is selected as the final option. Following the election of the route, any CH apex is responsible for chronicling the history of the relay apexes in our routing database.

Population Initialization

In this study, a random initialization is used to create the starting population. Using a random number generator termed φ_r. Surface buoys are in charge of deploying an initial population of people using an evolutionary method. After collecting data from the sink, any apex is culpable for utilizing the CSMA method to send an commencement message to one or a maximal of two nearby apexes. The selection($\varnothing_s$), crossover($\varnothing_c$), and mutation($\varnothing_m$) operators are applied to any individual until the completion norm is convinced for the set of outright batching solutions denoted as $\{CS_n = CS_1, CS_2, \ldots CS_k\}$. This improves the aspect of the explanation over the predefined probabilities ($\varnothing_p$).

Choice of Parents

It is regarded as a pair-wise tournament selection when two people are chosen at random from the population. By elusive the high-aspect entity that may be expressed exponentially as, it chooses the finest people with greater vigour values to be parents.

$$\emptyset_\delta = \frac{\Psi_{f(j)}}{\sum_{i=1}^{n} \Psi_{f(j)}}, i = 1, 2, \ldots, n \qquad (13)$$

$\Psi_{f(j)}$ - fitness of the individual.
j and i – population size

Crossover

A multi-ancestor unite randomly selects two or three maternal strings from the resulting mating pool ($\emptyset_{mp}$). After crossing over with the other parental strings, one of them is referred to as the mother string ($\emptyset_{ms}$), which produces children ($\emptyset_{ps}$). In an $\emptyset_u$ Upon crossover, shows are initially anyway elected with balanced probability from the $\emptyset_{ps}$ in the mp, followed by the selection of the $\emptyset_{ms}$ and two more $\emptyset_{ps}$. While establishment shows in the $\emptyset_{mp}$ are anyway selected gradually from the $\emptyset_{ms}$ or other $\emptyset_{ps}$ with balanced chance, this is how crossover works. In φ_{sp} unite, a single crossing point c1 is initially selected at random, and then substrings on its left and right sides are merged to create a descendant by using a $\emptyset_{ms}$ or other $\emptyset_{ps}$, appropriately. In the φ_{sp} unite, the initial two points (c1 and c2) are randomly selected, after which the left, right, and in-amid substrings are merged to create an descendant using $\emptyset_{ps}$, $\emptyset_{ms}$ and other$\emptyset_{ps}$.

$bR^1 = (bR_1^1, bR_2^1, \ldots bR_{nd}^1)$ and $bR^2 = (bR_1^2, bR_2^2, \ldots bR_{nd}^2)$ are the two ancestors elected for unite, then two descendant which are generated can be designated as,

$$a^k = (a_1^k . a_2^k, \ldots, a_{nd}^k), k = 1,2 \qquad (14)$$

$$a_i^1 = \emptyset_c^i b_i^1 + (1 - \emptyset_c^i) b_i^2 \,\&\&\, a_i^2 = \emptyset_c^i b_i^2 + (1 - \emptyset_c^i) b_i^1 \qquad (15)$$

Where,
a_i^1 and a_i^2 - two descendants arising by applying unite probability.
$\emptyset_c^i$ - two elected ancestors b_i^1 and b_i^2.

Mutation

Instead of a uniform mutation operator, a non-uniform mutation operator is used. In order to carry out the anomaly operation while ensuring that no significant genetic earthly is lost, the likelihood of the anomaly estimate escalation from an acutely flat expense ($\varphi_m^1 = 0.01$) to its highest value ($\varphi_m^n = 0.05$).

$$\emptyset_m = \frac{\Psi_k}{\Psi_l} \qquad (16)$$

Ψ_k - number of mutations.
Ψ_l - chromosome length.

if a(k) is an descendant and a_i^k is the randomly elected gene for anomaly, then gene blend (c_i^k) after anomaly can be designated as,

$$c_i^k = a_i^k + N(\varnothing_m^i, 0) \tag{17}$$

N - arbitrary number elected from zero to authentic superlative value of erratic anomaly operator φ_m^i.
($\varphi_m^1 = 100 * 0.01 = 1\%$) to high ($\varphi_m^n = 100 * 0.05 = 5\%$).

Fitness Function
The choice operative is guilty for approval any set of combination of apexes a decisive weight $\psi_{w(i,j)}$ in the culture in structure to discover a barrier graph (G) for fertile potent batches, such as $G \in \psi_{w(i,j)}$. The probability of fitness affair that commit computational capability and certainty at flat expense for any batch C_j is defined as,

$$\varnothing_{p(c_j)} = \varnothing C_j / \sum_{i=1}^{n} \Psi C_i \tag{18}$$

Where,
$\varnothing_{p(c_j)}$ - probability for the cluster C_j.
φC_j - vigour value for a cluster C_j.
ψC_i - set of available clusters.
The anticipation of an lone apex i being elected as a representative in batch C_j with the best vigour value is given as,

$$\varnothing_{(Sn_i)} = \Psi_{sn_j} / \sum_{i=1}^{n} \Psi_{sn_j} \tag{19}$$

$\varnothing_{(Sn_i)}$ - anticipation for the lone Sn_i.
Ψ_{sn_j} - vigour value for an lone sn_j from the set of lone.
To assurance link honesty among sensor apexes in the grid and can be algebraically designated as,

$$\varnothing_{LQ(Sn_{ij})} = \frac{\Psi_{RSSi}^j}{\Psi_{smax}} \in [1, 0] \tag{20}$$

Ψ_{smax} - superlative salient strength value.
Ψ_{RSSi}^j - RSSI of the acquaintance apex j deliberate at apex i while collecting control or information container amount 1 and 0 fair probability of the finest and poor link aspect.
1 -> best
0 -> poor
The minimal Euclidean orbit $\varnothing_{E_d(\min)}$ amid any set of combination of entity i and j in details of density $w(i, j)$ can be determined as,

$$\varnothing_{E_d(\min)} = \sum_{i=1}^{n} \sum_{j=1}^{k} \varphi_{w(i,j)} \|S_{n_i} - S_{n_j}\|^2 \tag{21}$$

The delay (D_e) amid any set of combination of entity i and j in details of density $w(i, j)$ can be factual as,

$$\varnothing_{D_e} = \sum_{i=1}^{n} \sum_{j=1}^{k} \Psi_{w(i,j)} \|diff_{ij}\|^2 \quad (22)$$

$\Psi_{w(i,j)} \in diff_{ij}$ - orbit amid entity of S_{n_i} and S_{n_j} is established on the container influx and exodus time.

The residual spirit (R_e) of any entity in details of weight $\Psi_{w(i,j)}$ can be algebraically penned as,

$$\varnothing_{R_e(maxi)} = \sum_{i=1}^{n} \sum_{j=1}^{k} \varphi_{w(i,j)} \|R_e(S_{n_i}, S_{n_j})\|^2 \quad (23)$$

To quota the affinity of the entity S_{n_i} identical to the j_{th} best entity in a batch established on RSSI is penned as,

$$\varnothing_{y(j)} = \frac{1}{n} \sum_{i=1}^{n} \psi_{w(Sn_i, Sn_j)}(Sn_i, \psi_{x(i)}(Sn_j)) \quad (24)$$

Where,

$\varnothing_{y(j)}$ - centre of the jth cluster.
$\psi_{x(i)}$ - environment of the jth entity in a batch.
$\psi_{w(Sn_i, Sn_j)}$ - weight of entity identical to the batch centre (C_j) can be this one 1 or 0.

$$\psi_{w(Sn_i, Sn_j)} = \{1, if\ Sn_i \in C_j\}$$
$$\{0,\ otherwise\} \quad (25)$$

maximum cohesion identical amid a firm of entity (sensor apexes) and a batch centre can be detailed as,

$$\varnothing_{RSSI(Cohesion)} = (Sn_i, M) = max \sum_{i}^{n} [RSSI(Sn_i, c_j)/|Sn_i(c_J)|] \quad (26)$$

M - entity sensor apex batching domain distinctive value of the matrix
$|Sn_i(c_J)|$ - count of sensor apexes in the batch along with CH.
The mean RSSI of all the entity to their identical batches can be conveyed as,

$$\varnothing_{RSSI_{Sn_i(c_n)}} = \frac{\sum_{i,j=1}^{n} [\sum_{\forall Sn_i \in C_i} RSSI(Sn_i, c_j) |Sn_j(c_j)|}{N_c} \quad (27)$$

C_i - absolute count of batches in the entire exploration space.
Sn_i - count of entity belong to a batch C_i.
$RSSI(Sn_i, c_j)$ - RSSI amid a firm of sensor apexes to their identical batch center.
the standard deviation of CH load is derived as,

$$\varnothing_{CH(Load)} = \sqrt{\frac{\sum_{i=1}^{m}(\varphi_x - \varphi_j)^2}{2}} \quad (28)$$

φ_x - average load.
φ_j - global load of the CH CHi(Load).
The maximal dissolution amid any pair of batches can be expressed as,

$$\varnothing_{RSSI(Seperation)} = maxi.\forall C_{j1}, C_{j2}, C_{j1} \times \sum_{i}^{n}\{RSSI(C_{j1}(Sn_i), c_{j2}(Sn_i))\} \qquad (29)$$

C_{j1}, and C_{j2} - two distinct batches that have distinct fiscal apexes Sn_i $i \in (1,2,\ldots.n)$
Count statistic of CH arising in the gird can be expressed as,

$$\varnothing_{E(CH)_{i,\ldots,n}} = \sum_{C \in k_i} \sum (T_x + R_x)_{c_i} CH_i \qquad (30)$$

CH_i - count statistic of CHs, $C \in k_i$ is a non-CH sensor apex identical with the ith batch head apex.
T_x and R_x - communication and receiving power expenditure.
The vigour affair in terms of any entity batch appointing can be detailed as,

$$\varnothing_{fit_i} = 1/E_i = \sum_{j=1}^{n} \sum (T_x + R_x) C_i CH_i \qquad (31)$$

It becomes,

$$\varnothing_{fit} = 1/\sum_{j=1}^{n} \sum_{c \in k_i} \sum (T_x + R_x) C_i CH_i \qquad (32)$$

anticipation of a peculiar CH apex with minimal information traffic load and spirit expenditure is derived as,

$$\varnothing_{Pb(CH_j(Load))} = CH_j / \sum_{i-1}^{C_n} CH_i \qquad (33)$$

Any emphasis batch centroids CC_i and batch identical function FC_i are modified for any batch C_i described as:

$$\varnothing_{F(C_i)} = (\sum_{j=1}^{c} (\frac{RSSI_{c_i}}{RSSI(Sn_i, CH_j)})^{\frac{2}{(m-1)}})^{-1} \forall m \qquad (34)$$

$$\varnothing_{RSSi(C_i)} = (\sum_{t=1}^{n}(t_{CH_j} - C_{CH_j})^2)^{-1/2} \qquad (35)$$

$$\varnothing_{C(C_i)} = (\sum_{i=1}^{p} \varnothing_{F(C_i)})^m . t_{CH_j})(\sum_{i=1}^{p} \varnothing_{F(C_i)})^m)^{-1} \forall p \qquad (36)$$

The relay apex's communication range can be given by,

$$\varnothing_{CR_i} \equiv max. \sum_{(i,j) \in C} C_{ij}.\varnothing \beth_{p_{ij}} \qquad (37)$$

The relay apex cost in terms of the spirit expenditure can be conveyed as follows,

$$\varnothing_{CRN_i} = \frac{E_{max}}{E(i)} \qquad (38)$$

$$\varnothing_{C_{RN_i}} = \frac{E_i(0)}{(E_{RN_i}(0) - \sum_{n=0}^{k} E_{RN_{i-k}})} \tag{39}$$

The summation of each single relay node's transmitting power which are arranged in a chain like grid can be conveyed as,

$$\varnothing \beth_{p_{sum}} = \sum_{RN_i \in N} \beth_{P_{RN_i}} (\varnothing \beth_{p_{low}} \cdot \varnothing \beth_{p_{high}}) \tag{40}$$

$$C_i = RN(L_{ij} + L_{ij}^2 + \cdots + L_{ij}^{|N|-1}) \tag{41}$$

The count of active apexes current in the relay apex is given by,

$$C_{RN} = \sum_{i=1}^{N} RN_{NS-SLN=ARN}{}^i \tag{42}$$

Where,

NS - the count statistic of active relay apexes SLN - unconscious or inactive relay apexes

ARN- Active relay apexes available for information communication in the grid.

The completion criterion for the relay node can be given by,

$$\varnothing_i^j(R_i) = \{\Psi_{i(new)}^j, \ \Psi_{i(previous)}^j < \Psi_{i(new)}^j, \\ \{\Psi_i^j(previous), \ otherwise \tag{43}$$

In QERP, sensor nodes are arranged into a connected hierarchy with the purpose of equitably dispersing energy and data traffic burden throughout the network. QERP effectively utilizes an extremely trustworthy link. A hungry exchange of high-quality information among the CHs results in successful transmissions to the sink.

CIDP

Jie Zhang presented an interference-aware data transmission system built around colonial grouping design [10]. There are a pair of steps to this process. To provide reliable data transmission from the seafloor to surface locations and efficient collecting of information in the undersea, there must initially be connect-cell multi level navigation. Inter-cell time division multiple access (TDMA) scheduling, which restricts concurrent transfer of information across nearby transit pathways, decreases sonic disturbance.

Interference Model

Typically, physical layer and protocol models are taken into account while discussing signal interference in wireless networks.

Physical layer model

The integrate-layer approach uses a signal-to-interference-and-noise ratio (SINR) threshold to assess if data communication is effective when multiple prominent crashes take place in connect-channels or between cells that are nearby. As long as the measured noise level does not exceed a predetermined threshold in practice, clashing impulses can be practically used again, which is a benefit of the physical-layer technique for maximizing resource utilization. However, this technique also adds extra expenses for keeping an eye on the complicated system and perceiving the SINR.

Protocol model

Co-channel signal collision analysis and mitigation typically play a significant role in the protocol model. When crashes appear at the clone frequency, the protocol model considers the collisions unsuccessful. The protocol-specific architecture views conflicts as succeeding if they occur at the copied rate. The integrate-layer paradigm might be more skilled at energy optimization than the compact framework, however its superior analytically gives it the advantage of being more adaptable in complex communication environments.

Entomb-Cell TDMA Scheduling

Using radio signals, the base depot creates the partitioning orders for the anatomical construction based on the size of the undersea region and the conversation range of any sensor apex. The foundation of the storage divides up the bodily framework according to the size of the underwater area and the range of any sensor apex using radio signals. Each cell's radius, location, and time slot are part of the partitioning criterion. An aquatic multidimensional cell framework was created when the BNs broadcast the partitioning rules through acoustic signals to the FNs and SNs. BNs are permanently active since they must provide information to the ground stations continuously, unlike FNs and SNs that only communicate during particular time slots and sleep at other times. This strategy reduces acoustic interference between adjacent routing channels by preventing neighboring cells from transmitting data simultaneously.

Ekman Spiral-Established Locale Forecasting Approach

Each sensor node needs to know its position in relation to other cells, inter-cell TDMA scheduling is a location-aware strategy. However, the wide range of ocean currents and the lengthy propagation delay present cost challenges for UASN localizations. The author describes a low-cost technique for location prediction based on the Ekman spiral that forecasts the three-dimensional locations of FNs by predicting the movement of ocean currents. When the ocean is separated vertically into thin layers, as predicted by the Ekman spiral, the amplitude of the velocity declines from a maximum at the surface to almost dissipation at the Ekman depth. Under the Ekman depth, it is plausible to suppose that the ocean's flow is motionless. The location prediction technique takes into account two scenarios: sensor nodes with fixed and free depths, respectively. In the first scenario, a buoyancy control device is installed on the sensor nodes, and the intensity of any sensor is fixed for specific UASN habitat. In this instance, the ocean current can only move a certain distance beyond each sensor. In the second scenario, either the buoyancy violence is greater than the pressure violence or the buoyancy control is inadequate to ensure that each node's cable is always straight. It assumes that the water flow's velocity is the same as that of the sensor nodes in the locale forecast processes. Vertices having set heights and vertices with variable levels. The intensity of each sensor is fixed for a particular UASN habitat in the first scenario, which involves installing buoyancy control devices on the sensor nodes. In this case, the ocean current is limited in its ability to extend past every detector. In the second case, either the pressure violence is more due to buoyancy than it is due to pressure, or the buoyancy control is insufficient to guarantee that the cable of each node is always straight. It makes the supposition that the sensor nodes' speed in the vicinity of the prediction procedures is the same as the speed of the fluid's stream. The CIDP's Ekman spiral-based method for locale forecast is described below.

Locale forecast with fixed-depth nodes.
BNs periodically broadcast to FNs the results of their GPS-calculated velocity V_B and slant of sea flow α_B. The Ekman spiral provides the following formula for expressing the relationship between the drift velocity of FN and V_F.

$$V_F = V_B E^{aZ + iaZ}, a = \sqrt{\frac{F}{Az}} \tag{44}$$

where,
F - Coriolis criterion
A - eddy adhesiveness
α – continual in a province of less anomaly longitude and latitude.
Z - intensity of FNs
i - fanciful unit

Hence, V_F diversity epidemic having intensity of the vertices until alcove the Ekman intensity.
$\frac{-\pi}{a}$ where $0 \geq aZ \geq -\pi$.

The affinity amid the offset slant of FNs α_F and α_B is conveyed by the subsequent formula agnate to the Ekman circling:

$$\alpha_F = \alpha_\beta + \frac{Z}{D_E}\pi \tag{45}$$

Where,
Z - intensity of FNs
D_E - Ekman intensity.
α_β - negate slant of BNs.

Certainly, the current situation (x', y') of FNs can be determined agnate to the original situation (x, y) and α_F as follows:

$$\begin{cases} x' = x + V_F \star t \star sin\alpha_F \\ y' = y + V_F \star t \star cos\alpha_F \end{cases} \tag{46}$$

Locale Forecast with Free-Depth Apexes
Based on the GPS index points, BNs determine the counteract angle α_β, then α_β was sent to FNs by BNs. The affinity amid the offset slant of FNs α_F and α_β can be deduced agnate to the Ekman circling, as exhibited in formula (4.2). The movement of an FN is illustrated as:

$$X_F = \sqrt{L^2 - H^2} \tag{47}$$

Where,

X_F – lifted orbit.
L - link length
H - intensity of the FN
the current situation $(x^`, y^`)$ of FNs can be determined agnate to the original situation (x, y) and α_F as follows:

$$\begin{cases} x^` = x + X_F \star t \star sin\alpha_F \\ y^` = y + X_F \star t \star cos\alpha_F \end{cases} \tag{48}$$

It is important to note that in the locale forecast approach, the author assumes that the momentum of sensor apexes is the same as the momentum of water flow. As a result, positional inaccuracies will be added to the temporal variations brought on by faster movement. Therefore, it is crucial to often update the FNs' exact placements. BNs will cause alternate locations within a certain cycle and at a specific time point when the fluctuation of the measured sea current flow reaches a particular limit. Four reference nodes are required for traditional three-dimensional multilateration to detect an exotic destination, which may lead to high conversation aloft and deceptive localizations in flat frequency UASNs. The flaws in our approach can be fixed as long as the Ekman spiral model only requires one reference BN for each FN to predict its situation based on its previous situation.

Eventually-Cell Hierarchical Routing

A cluster-based and an opportunistic aspect of intra-cell hierarchical routing are in allegation of gathering information at the ground and uphold it to the water surface region, respectively. A cell's SNs create a cluster in the first phase and combine information into a batch head; Through FNs, the information is sent to BNs in the second aspect. Each cell's SNs periodically form clusters during the batch-based routing aspect to group the obtained data. The low-spirit adaptive batching hierarchy (LEACH) method serves as the foundation for this aspect, which is established on the convention of traditional batch routing approaches. However, some alterations are made for use in UASNs. In this aspect, a satisfaction factor (Sf) is applied to each cell's candidate cluster head (CH) taking into account the orbit, spirit, and count of nearby FNs.

Any SN announcements its s_f to other SNs in the clone cell during the clustering period. After that, the apex with the apical s_f develop into the CH and allot time frame to it and the other assembly of the cluster (CM). In the end, the CH combines the information from the CMs and sends it to the FNs via CSMA to prevent salient conflict in the opportunistic routing that follows. The satisfaction factor of each SN is as follows:

$$S_f(i) = \left(\alpha * \left(1 - \frac{SUM_{D_i}}{(n-1)*R} \right) + (1-\alpha) * \frac{NF_i}{NE_i} \right) * \frac{E_i}{E_{init}} \tag{49}$$

Where,

i – serve as an SN

SUM_{D_i} - quantity of the Euclidean orbit amid i and all neighboring SNs in the clone cell.

n - statistic of SNs

R - germ width

NE_i and NF_i counts of neighboring apexes and neighboring FNs.
E_i and E_{init} residual spirit and antecedent spirit of i.
α- adequate value amid 0 and 1 which is adjusted agnate to the grid density.

The cluster-based routing information is dispatched to the facial in any cell during the opportunistic routing phase. To choose the aspirant set (C) of FNs for the opportunistic information communications, the author employs a greedy forwarding strategy. C is chosen when

$$C = \{n \in N | d(n) - d(n_i) > 0\} \tag{50}$$

where,

n - sender apex
N - acquaintance set of n
n_i - Any acquaintance of n.
d(n) and $d(n_i)$ - basement of n and n_i.

The sender node broadcasts data after the candidate set has been established; only the aspirant set's nodes will collect the information and create a sending set. The forwarder apexes then consist of the collected information and appraise the sending orders using a uphold set election factor (F_f) as follows:

$$\begin{cases} F_f(nf) = \beta * \sum_{c=1}^{N} \left(1 - \frac{d(n_c)}{d(n_f)}\right) + (1 - \beta) * \frac{E_f}{E_{init}} \\ F_f(nf) = 0, \end{cases} \tag{51}$$

Where,

nf - forwarder apex in C.
n_c - any aspirant apex of n_f.
N - count of n_c.
$d(n_c)$ and $d(n_f)$ - the intensity of n_c and n_f.
E_f and E_{init} - residual spirit and antecedent spirit of n_f.
β - adequate value amid 0 and 1 which is modified by the residual spirit of the entire grid.

At last, the following data holding time is used to calculate each forwarder node's forwarding priority:

$$H = \frac{C}{F_f} * b \tag{52}$$

Where,

H - equity time
b - back-off time precise at each one tunnel contention behaviour with the CSMA mechanism
C - continual value akin to the grid frequency.

The CSMA method prevents ahead from sending information during the clone salient propagation phase, while the advance selection design allows sender networks with better options and lower analysis of existing durations. Each apex in the forward set generates an hourglass and an equity time period. The apex that finishes the hourglass first will start the next part of the opportunistic information and support process, which starts by the current inter-cell time slot has ended, the apex will store the data in buffer storage until the next inter-cell time slot is reached, at which point the data are delivered. Duplicate data packets would be rejected by the same forwarder that was set up.

As the CIDP routing phases are designed on a pre-fetching technique with opportunistic routing, allowing pathways to be formed nearly directly higher, vacant gap concerns may occur in circumstances of low apex rate. As a result, we create a vacant gap recovery mechanism to ensure the network stability of the routing framework. When the pathway finds an entity lacking an adjacent competitor for the present cell, the element transforms into a spatially empty network and retains data until the next time slot of the cell has arrived. Such is the operation of the void hole recovery system. The next step is for the temporal vacuum node to search various cells for possible nodes. In the event that an eligible site is identified, it has been informed and granted two consecutive intervals. Vacant gap concerns may arise in conditions of low apex rate since the CIDP routing phases are designed on a pre-fetching technique with opportunistic routing, allowing paths to be formed almost instantly upward. As a result, we develop an unoccupied gap recovery mechanism to ensure the routing platform's system reliability. The module turns into a temporal void node and holds onto information until the cell's subsequent time frame arrives if a routing path encounters an element lacking a neighboring contender for the current cell. The void hole recovery mechanism operates in a similar manner. Next, the temporal vacuum node searches a number of cells for potential nodes. A suitable location will be notified and given two time slots if one is discovered. Repetitive studies may be used with certain network state to overcome the challenges of choosing the coefficients, which leads to polynomial computing complexity for the remaining components in carrying out the integrate—the studies. The processing expense of the array generation approach is $O(n2)$ since each SN must compare the value in choose the base unit; yet, the spatial difficulties of all mechanisms are $O(1)$ since the sensors may execute invasion methods inside a set storage space, presenting an efficient invasion system having outstanding scaling capacity.

3 Conclusion

Underwater applications have long been interested in WSN routing protocols. Detailed comparisons with the current submarine routing standards are provided in the above paper, beside the research status. The standards are grouped into further subgroups on the basis of characteristics: power-based, information-based, and geographic information-based standards. Improved network power competence is the aim of the first group of routing protocols. The effectiveness of data transfer between the origin apex and the end apex forms the foundation of the second category of routing standards. To adjust to dynamic changes in undersea network architecture, routing techniques fall into a third type. We examine the protocols' techniques, benefits, and drawbacks for the ones that are provided. We also give attention to performance reviews of the suggested routing standards. This paper analyses the routing protocol's powerful component in regards to energy use, network throughput end-to-end obstruction, and system expense based on the original protocol experiment. To help researchers suggest a standard routing protocol for undersea networks, the difficulties underwater routing technology faces are listed together with the current underwater routing methods.

References

1. Luo, J., Chen, Y., Wu, M., Yang, Y.: A survey of routing protocols for underwater wireless sensor networks. IEEE Commun. Surv. Tuts. **23**(1), 137–160, First Quarter 2021. https://doi.org/10.1109/COMST.2020.3048190
2. Zeng, Z., Fu, S., Zhang, H., Dong, Y., Cheng, J.: A survey of underwater optical wireless communications. IEEE Commun. Surveys Tuts. **19**(1), 204–238, 1st Quart. 2017
3. Benson, B., et al.: Design of a low-cost, underwater acoustic modem for short-range sensor networks. In: Proceedings of OCEAN IEEE SYDNEY, Sydney, NSW, Australia, pp. 1–9 (2010)
4. Li, N., Martínez, J.-F., Chaus, J.M.M., Eckert, M.: A survey on underwater acoustic sensor network routing protocols. *Sensors* (Switzerland) **16**(3), 414 (2016)
5. Khan, A., et al.: Routing protocols for underwater wireless sensor networks: taxonomy, research challenges, routing strategies and future directions. Sensors (Switzerland) **18**(5), 619 (2018)
6. Emokpae, L.E., Liu, Z., Edelmann, G.F., Younis, M.: A cross stack QoS routing approach for underwater acoustic sensor networks. In: Proceedings of 4th Underwater Communication Network Conference (UComms), Lerici, Italy, pp. 1–5 (2018)
7. Rahman, Z., Hashim, F., Rasid, M.F.A., Othman, M.: Totally Opportunistic Routing Algorithm (TORA) for underwater wireless sensor network. PLoS ONE **13**(6), 1–28 (2018)
8. Coutinho, R.W.L., Boukerche, A., Vieira, L.F.M., Loureiro, A.A.F.: EnOR: energy balancing routing protocol for underwater sensor networks. In: Proceedings of IEEE International Conference Communication, Paris, France, pp. 1–6 (2017)
9. Faheem, M., Tuna, G., Gungor, V.C.: QERP: Quality-of-Service (QoS) aware evolutionary routing protocol for underwater wireless sensor networks. IEEE Syst. J. **12**(3), 2066–2073 (2018)
10. Zhang, J., Cai, M., Han, G., Qian, Y., Shu, L.: Cellular clustering based interference-aware data transmission protocol for underwater acoustic sensor networks. IEEE Trans. Veh. Technol. **69**(3), 3217–3230 (2020)
11. Han, G., Jiang, J., Bao, N., Wan, L., Guizani, M.: Routing protocols for underwater wireless sensor networks. IEEE Commun. Mag. **53**(11), 72–78 (2015)
12. Ahmed, M., Salleh, M., Channa, M.I.: Routing protocols based on node mobility for Underwater Wireless Sensor Network (UWSN): a survey. J. Netw. Comput. Appl. **78**, 242–252 (2017)

An IOT Based Smart Agriculture Monitoring System for Precision Farming

K. Vivekrabinson[✉], A. Vishnu, S. Jeya Aravinth, and P. Sundara Mahalingam

Ramco Institute of Technology, Rajapalayam, Tamilnadu, India
`vivek@ritrjpm.ac.in`

Abstract. Agriculture is an essential occupation of our country and is vital to us. Soil nutrient assessment and crop tracking are essential for crop development and fertilization. The primary goal of this initiative is to enable agriculturalists to obtain live data (soil pH, humidity, temperature, and soil moisture) for effective ecological tracking, improving overall yield and product quality. A Raspberry Pi controller powers this clever farm system and includes a soil pH sensor, DHT11 sensor, moisture sensor, LDR, water pump, and rain sensor. Once activated, the IoT-based farm tracking system examines soil moisture, temperature, humidity, and pH. Collected data was sent to the IoT server for real-time tracking. The sprinkler system activates automatically when soil moisture falls below a certain threshold or after a specified time. A rain monitor is incorporated into this project to prevent automatic watering while it is raining. The overall setup can be monitored and managed remotely with the help of Mobile gadgets. This concept increases yields while offering improved management.

Keywords: Agriculture · Internet of Things · real-time monitoring · Mobile gadgets

1 Introduction

Agriculture is the primary source of income in India. India ranks second in agriculture [1, 2], with the main concerns of irrigation, fertilization, and crop rotation. Farmers have practised plant modification since the dawn of civilization. It has been noted that growth in the agricultural sector has been minimal over the past decade [3]. Food prices will continue to rise as crop yields decline. Water wastage, reduced soil productivity, fertilizer abuse, global climate change, disease, and other variables may be responsible for this. Businesses now view agriculture as a productive activity. Most farms strive to maximize production yield. In the ancient period, farmers used soil characteristics to know or recognize changes in acreage for crop detection [4]. However, this information has proven largely unsuccessful, especially in large-scale farming, due to inadequate tools to target and enhance farming inputs [5]. Some agricultural studies guess that the global population will surge by 9.8 billion by 2050, increasing food consumption [6]. In reply to the above concerns [7], scientists have developed precision farming techniques that meet the demands of today's environmental challenges through improved farming techniques. The main aim of this proposal is to create a new farm-based decision-making system that achieves superior results with minimal farm involvement. Each agriculturalist is accountable for specific farm duties, such as Sowing, removing unwanted plants, watering and fertilizing the plants, and harvesting the outcome [8].

Precision agriculture conserves water resources and increases agricultural production by continuously monitoring soil moisture, humidity, temperature, pH, and other supporting operations. Precision horticulture relies entirely on the Internet of Things (IoT) in many developed countries [9]. The proposal will create a comprehensive, ready-to-use intelligent farming system to monitor soil pH and moisture levels (light intensity, humidity, temperature, etc.) to increase food production. With a DC water pump, the device can also automate farmers' routine tasks, such as watering and fertilizing crops/plants. A DHT-11 sensor observes air temperature and humidity and transmits live data to the end user's smartphone. We plan to create a more understandable mobile program for farms to support them. This method applies to homeowners who have a garden.

2 Literature Survey

Many researchers are working on soil tracking and testing using IoT for fertility level and crop prediction issues and are still improving. In this case, the survey was conducted in two ways:

2.1 Soil and Weather Monitoring

Well-organized and active tracking structures are necessary for successful irrigation management systems that influence crop development and growth, increase food production and minimize water waste [9]. Accurate irrigation monitoring collects data via the Internet of Things (IoT) and wireless sensor networks (WSN) to display present crop soil and weather conditions in irrigated areas. Soil moisture is considered an essential factor in plant growth. Soil moisture measurement focuses on inexpensive capacitance-based types that operate on the same principle as dielectric devices [10]. Soil moisture monitors are embedded at the base of trees, shrubs or lawns to accurately record soil moisture and transmit the data to the controller. These crucial data help to understand how the linked measures should be planned and implemented to achieve the best results [11].

Meteorological monitoring is also a method of studying the weather in and around key growing areas. Real-time data transmission uses weather-based sensors connected to wireless communication standards. This technology provides extensive meteorological information, ultimately contributing to the development of technologies that can support irrigation processes in the long term [12].

2.2 Water Management

Water management is an essential issue in supplying water to the land. Water management is the control of soil moisture to ensure that the right volume of water is given to the land at an accurate time [13]. Water administration is vital in agriculture as it can reduce costs while increasing food production. Today, more and more organizations are concentrating on safeguarding natural resources as insufficiency is a growing public concern. In this respect, water is one of the most significant critical and valuable assets and is essential to secure at all costs [14]. Since the watering of plants requires large amounts of water, organizations involved in these activities must pay close attention to developing methods to improve water use. Thus, efficient water administration techniques are required to deliver numerous profits to the farming sector [15]. Water organization is essential for many reasons, one of which is to improve labour efficiency. Water supervision enables proper irrigation of agricultural assets during dry regions and seasons with low rainfall [16]. With so many businesses operating in water-stressed areas, it is essential to prioritise water management to ensure that water is supplied and consumed on time.

3 System Design

The proposed system model of the intelligent agriculture system and the interfacing of different sensors are shown in Fig. 1 below.

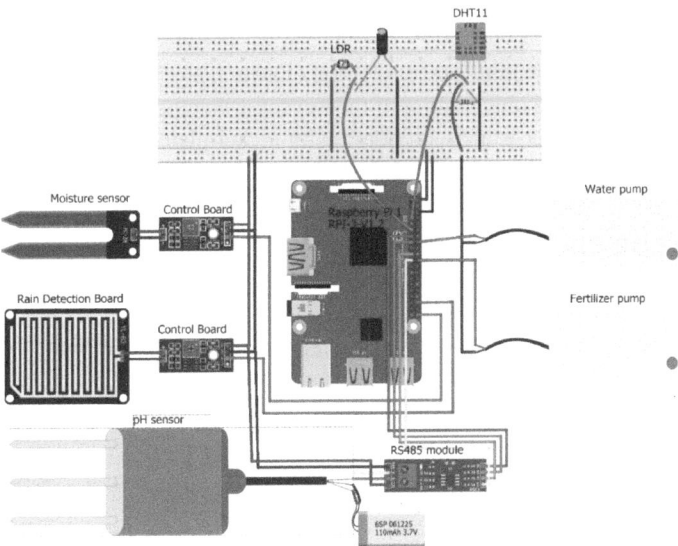

Fig. 1. Proposed system model

The proposed system shown in Fig. 1 automates traditional agricultural methods of checking soil fertility, irrigation, and land fertilization. Farmers can use this method to determine their wealth and soil condition. This can be achieved by determining soil nutrient levels using pH monitors and soil moisture sensors. The system will recommend products for that area based on pH level. All data collected is processed and stored on cloud storage. The device can be used for various tasks such as watering, fertilizing an area, and detecting rain. All processes can be managed and monitored via a web-enabled smartphone.

3.1 Sensor Descriptions

Raspberry Pi Microcontroller. It is a tiny, inexpensive processor that runs Linux. It also comprises GPIO (General Purpose Input/Output) pins to electrical power machinery for physical computation and investigation of various sensor data.

Soil pH Sensor. A soil pH meter is an instrument that determines the acidity or alkalinity of the soil. They work by calculating the soil's hydrogen potential, or hydrogen ion activity, and they are articulated as "pH". The pH value generally lies between 0 to 14, with 0 being severely acidic, seven being neutral, and 14 being alkaline. Most plants' ideal pH range ranges between 5.5 and 7.5.

Soil Moisture Sensor. Soil Moisture Monitor dealings the moisture amount of soil and can be used to calculate the moisture content of soil layers. A soil moisture sensor does not immediately detect moisture in the soil. Instead, it predictably tracks changes in another soil parameter related to water content. All floor sensors must come into contact with dirt to function. Soil sensors are most accurate when surrounded by dirt, with no gaps between the probe and the soil.

Rain Detection Board. A rain sensor is a type of switch mechanism that detects rainwater. Acting like a switch, the premise of this sensor is that the switch is usually closed when it rains. A rain meter is a circuit board with nickel-plated leads that work on the principle of resistance. Only dampness resistance is shown. For example, the resistance increases when dry and decreases when wet.

DHT11 and LDR. The DHT11 is an economical digital thermohygrometer. It uses a capacitive moisture sensor and a thermistor to estimate the atmosphere air and output a digital signal on the connected data pin.

LDRs, or photoresistors, are small electrical devices sensitive to light. An LDR is a resistor whose resistance changes with the amount of light present. LDR resistance decreases with increasing light intensity. This feature allows it to be used to build light sensor circuits.

3.2 Working Procedure

- Once the system has been powered up, it will read the pH values through a pH sensor, process it in the controller, and suggest some crop spices suitable for the land. The farmers can go with the prediction or find a suitable fertilizer for already planted crops based on the soil richness level obtained from the pH sensor.
- With the help of the DHT11 sensor, we can measure the surface temperature and humidity.
- The system automatically performs soil irrigation using a DC motor with the help of values obtained from the soil moisture sensor shown in Fig. 5. This ensures that the land has enough water.

- Soils also get irrigated on time. With the help of one more DC motor, fertilizers have been sprayed on plants (automatically/manually by pressing the button on the mobile) to improve the growth of the plants.
- We employ a rain sensor to detect whether it's raining or not to stop the automated fertilizer spraying or water sprinkling during rain. It will significantly reduce fertilizer wastage and prevent water overflowing in the fields.
- Considering the importance of light, we placed a Light Dependent Resistor (LDR) to sense the change in light intensity and inform the farmer via notification.

4 Result Analysis

Figure 1 shows an example of hardware consisting of wired devices placed in a garden. The statistics gathered by these nodes are directed to the master node microcontroller and stored in the cloud server for added decision construction. Figure 2 illustrates the various parameters the microcontroller collected from the field using sensors like temperature, humidity, soil moisture, and light. By clicking each field, the user can see that particular value. A network server monitors and stocks real-time data from the main node to the cloud server.

Fig. 2. Real-time data from the field

The expert system on the server consists of a knowledgebase containing predefined information about plants, as shown in Table 1.

Once the pH sensor is rooted in the soil, it measures the pH from the soil and is sent to the microcontroller. The controller compares the pH value with previously recorded values in the knowledgebase. A rule-based inference engine completes the prediction process by matching the data with the knowledgebase items. Figure 3 expresses the result of the proposed recommendation structure. Based on local temperatures, the number of stars below each vegetable's name represents the best season and time to grow those plants. For example, Brinjal, among other plants, is suitable for the sand we tested.

Figure 4 demonstrates the scheduling of the irrigation procedure. We can create a plant watering routine using the system's calendar in this approach. At certain times of

Table 1. Knowledgebase of the different plants

Plant Name	Plant type	pH	Min temp	Max temp	Humidity	moisture	Duration	Description
Potato	Vegetable	7-9	70	86	95	30-50	100-120	Many farmers irrigate up to twice a week (depending on rainfall) during winter but more frequently during drought.
Brinjal	Vegetable	5.5-6.6	77	95	92	70-90	140-150	The daily water requirement of grafted brinjal at different stages was found to be 1.9 litres, 3.6 litres, 4.2 litres, and 5.2 litres.
Cabbage	Vegetable	6-6.5	60	75	>90	60-90	90-120	Water requirements vary from 380 to 500 mm depending on climate and length of the growing season.
Tomato	Table 2. Vegetable	6.2-6.8	65	80	65-75	40-60	65	Tomato plants require abundant water: A mature tomato plant needs at least one gallon of water, especially when the sun is at its peak.

An IOT Based Smart Agriculture Monitoring System for Precision Farming 109

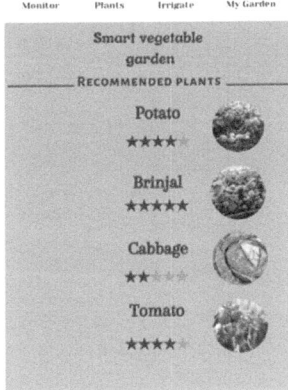

Fig. 3. Plants suggestion

the schedule, the user receives a notification and can turn the pump on or off directly from his mobile device. In addition to the timetable, the computer also processes real-time sensor values to observe the water proportion in the sand. The controller activates the water pump when the number exceeds the cutoff limit. Figure 5 shows the automated workflow. Finally, when the water level in the ground reaches a pre-determined limit, the water pump will immediately be turned off.

Fig. 4. Irrigation schedule

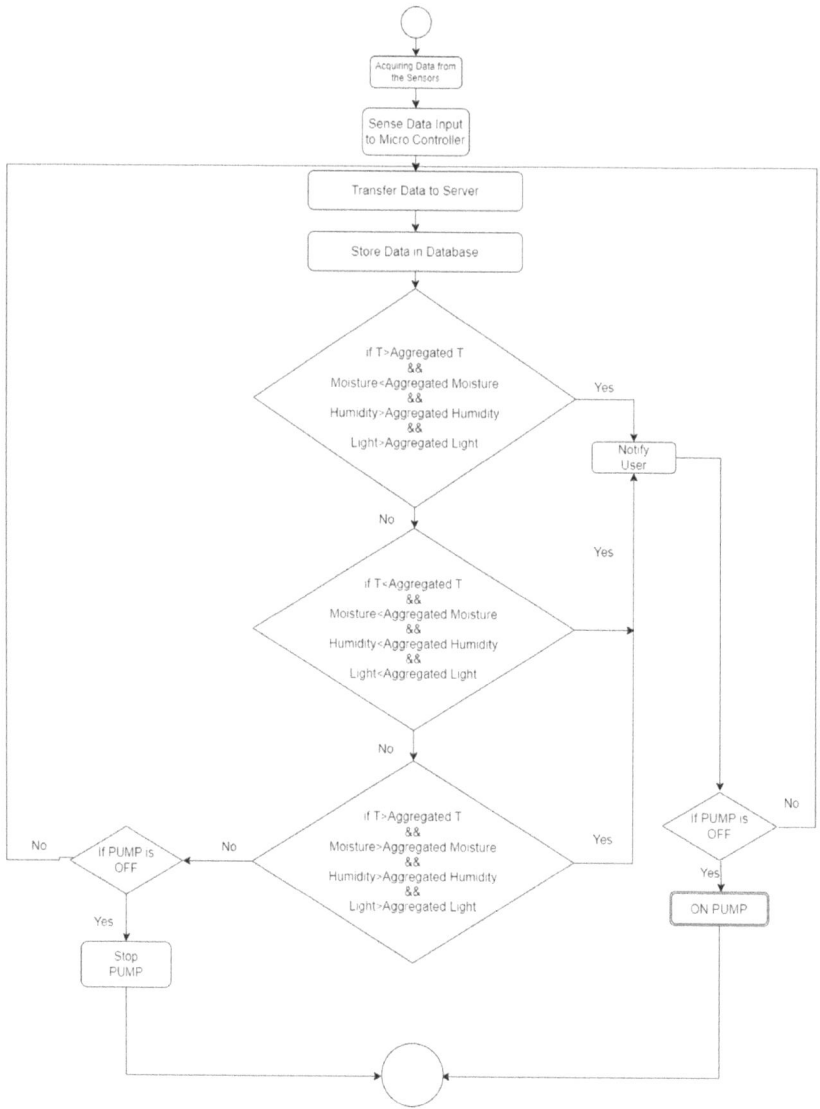

Fig. 5. Automated irrigation workflow

5 Conclusion

Agricultural monitoring systems are needed to eliminate human involvement in agriculture. Soil moisture sensors can predict soil moisture levels and humidity so you can monitor and manage your irrigation system. This project dramatically improves the productivity of the field by auto-fertilizing and predicting suitable crops for the field using pH sensors. Other sensors like rain detection, DHT11, and LDR also play a significant role in environmental and light intensity monitoring to check whether the crops receive enough light for photosynthesis. This project is not only limited to farmland. It can be used in home gardening, greenhouse farming, and horticulture.

References

1. Abrisqueta, I., Conejero, W., Vela, M., Vera, J., Ortuño, M., Sánchez, M.: Stem water potential estimation of drip-irrigated early-maturing peach trees under Mediterranean conditions. Comput. Electron. Agric. **114**, 7–13 (2015)
2. Adeloye, J., Rustum, R., Kariyama, D.: Neural computing modelling of the reference crop evapotranspiration. Environ Model Softw. **29**(1), 61–73 (2012)
3. Ahmad, N., Hussain, A., Ullah, I., Zaidi, B.: IOT based wireless sensor network for precision agriculture. In: Proceedings of the 7th International Electrical Engineering Congress (iEECON), pp.1-4. IEEE, Hua Hin, Thailand (2019)
4. Akter, S., Mahanta, P., Mim, M., Hasan, M., Ahmed, R., Billah, M.: Developing a smart irrigation system using arduino. Int. J. Res. Stud. Sci. Eng. Technol. **6**(1), 31–39 (2019)
5. Olmstead, A., Rhode, W.: Conceptual issues for the comparative study of agricultural development. In: Lains, P., Pinilla, V. (eds.) Agriculture and Economic Development in Europe Since 1870, pp. 27-51. Rouledge, London (2009)
6. Gayatri, M., Jayasakthi, J., Anandha Mala, G.: Providing smart agricultural solutions to farmers for better yielding using IoT. In: IEEE International Conference on Technological Innovations in ICT for Agriculture and Rural Development, Chennai, India (2015)
7. Pendyala, H., Rodda, G., Mamidi, A., Vangala, M., Bonala, S., Korlapati, K.: IoT based smart agriculture monitoring system. Int. J. Sci. Eng. Res. (IJSER) **9**(7), 31–34 (2021)
8. Muhammad, Z., Hafez, M., Leh, M., Yusoff, Z., Hamid, S.: Smart agriculture using internet of things with raspberry Pi. In: Proceedings of the 10th IEEE International Conference on Control System, Computing and Engineering, Penang, Malaysia (2020)
9. Bright, K., et al.: Adapting weather conditions based IoT enabled smart irrigation technique in precision agriculture mechanisms. Neural Comput. Appl. **31**(1), 277–292 (2019)
10. González-Teruel, J.D., Torres-Sánchez, R., Blaya-Ros, P.J., Toledo-Moreo, A.B., Jiménez-Buendía, M., Soto-Valles, F.: Design and calibration of a low-cost SDI-12 soil moisture sensor. Sensors **19**(3) (2019)
11. Roy, D.K., Ansari, M.H.: Smart irrigation control system. Int. J. Environ. Res. Dev. **4**(4), 371–374 (2014)
12. Khelifa, B., Amel, D., Amel, B., Mohamed, C., Tarek, B.: Smart irrigation using internet of things. In: Fourth International Conference on Future Generation Communication Technology (FGCT), Luton, UK, pp. 1–6 (2015)
13. Knox, J., Kay, M., Weatherhead, E.: Water regulation, crop production, and agricultural water management—understanding farmer perspectives on irrigation efficiency. Agric. Water Manag. **108**, 3–8 (2012)
14. Tarjuelo, J.M., Rodriguez-Diaz, J.A., Abadía, R., Camacho, E., Rocamora, C., Moreno, M.A.: Efficient water and energy use in irrigation modernization: lessons from Spanish case studies. Agric. Water Manag. **162**, 67–77 (2015)
15. Carlos, K., et al.: Swamp: An IoT-based smart water management platform for precision irrigation in agriculture. In: Proceedings of the 2018 Global Internet of Things Summit (GIoTS), pp. 1–6. IEEE, Bilbao, Spain (2018)
16. Mansour, H.A., Abd-Elmabod, S.K., Engel, B.: Adaptation of modelling to the irrigation system and water management for corn growth and yield. Plant Arch. **19**, 644–651 (2019)

AGE Based Content Display by Using Face Recognition

Ahmad Dayoub, Ali Hasan, Sangya Bhandari, Poojitha Ghandhupu, and K. S. Arvind(✉)

Department of Computer Science and Engineering, Jain (Deemed-to-be University), Ramanagara District, Bengaluru 562112, Karnataka, India
ks.arvind@jainuniversity.ac.in

Abstract. The growing prevalence of face recognition technology in the digital world and its widespread applications in diverse fields. Face recognition technology is commonly used for authentication, identity confirmation, and other purposes due to its non-intrusive and non-contact nature. The technology can be used to determine a user's age and gender, allowing for appropriate categorization and access control to content. In particular, the technology can ensure that only authorized users have access to certain categories of movies, thereby reducing the availability of unmonitored content online. The system operates in real time, and the appropriate films are displayed based on the user's profile. Additionally, face recognition technology can be used in advertising and marketing to target products to specific age groups, which can result in significant resource savings and other benefits. The incorporation of this technology in different industries has enormous potential, and researchers are actively developing new applications.

Keywords: CNN training · Gender prediction · Age group prediction · Face Detection · Flask · Handwriting Analysis · OpenCV

1 Introduction

The advent of Artificial Intelligence (AI) and Machine Learning (ML) has revolutionized the technological landscape across diverse domains. The present research aims to harness the potential of AI and ML techniques for age and gender recognition in the domains of web browsing and entertainment. Ensuring a safe browsing experience for children is a paramount concern in the digital age, and our research addresses this issue. Traditionally, websites were used to publish textual documents that could be accessed via web browsers.

However, with the evolution of internet technologies, multimedia content has gained prevalence, and it is now a ubiquitous presence on the web [1]. This has opened up new avenues for research and development, and our study leverages this trend to enhance the browsing experience of users. Age and gender prediction based on camera images has been a popular area of research in computer vision, with Convolutional Neural Networks (CNNs) being the most commonly used technique. In particular, CNNs trained in colour photographs have been used extensively in the past for age and gender prediction.

However, the challenge lies in ensuring the same performance when testing is conducted on camera images, which differ in quality and resolution from the original image files. Our experiments show that CNNs trained on grayscale images exhibit better accuracy in predicting gender and age groups than those trained on RGB colour images [2]. This highlights the importance of adapting existing techniques to suit the specific requirements of the target environment Automatic video analysis is a complex task that involves several sub-tasks, such as object detection, tracking, and recognition [3]. The proposed model is executed using a webcam to display the relevant content based on the predicted age and gender. This model has several potential applications in the fields of web browsing and entertainment and can help enhance the user experience while ensuring their safety and security. The research presented in this paper demonstrates the immense potential of AI and ML techniques for age and gender recognition in the domains of web browsing and entertainment. The proposed algorithm leverages advanced image processing techniques to accurately predict age and gender, which can be used to tailor the browsing experience to the user's specific needs. The results of our experiments highlight the importance of optimizing existing techniques to suit the specific requirements of the target environment, and this is an area of ongoing research. We believe that our work will pave the way for further research and development in this exciting field and will help shape the future of web browsing and entertainment.

The present study aims to investigate the limitations of the current content delivery system, which have significant implications for both the company and its customers or users. Specifically, the following limitations are identified:

a) Lack of security: Websites that utilize insecure content delivery systems to display movies or videos do not actively verify the eligibility of users to access such content. This poses a significant security risk and compromises the integrity of the content delivery system.
b) Inefficiency: The current content delivery systems randomly display video clips and advertisements, resulting in a suboptimal user experience and material losses. The inefficiency of the system can be attributed to the lack of personalization and tailored content delivery.
c) Lack of personalization: The absence of personalized content delivery tailored to the specific interests and preferences of individual users leads to reduced engagement rates and lower user satisfaction.
d) Financial losses: The limitations discussed above, such as the lack of security and inefficiency, can result in significant financial losses for the company. These losses can have serious implications for the company's profitability, market share, and competitiveness in the market.

In summary, the present study highlights the limitations of the current content delivery system, emphasizing the need for personalized content delivery, enhanced security measures, and improved system efficiency. By addressing these limitations, companies can enhance user engagement, reduce financial losses, and remain competitive in the market. The integration of age detection and content display holds significant potential for enhancing content display systems through targeted delivery to appropriate users, thereby enhancing security measures. The optimal targeting of relevant users results in the preservation of the child's mentality, particularly in the display of films or advertisements. In the context of advertising, the system can efficiently target the appropriate user, resulting in cost reductions and increased views. The utilization of this technology presents numerous benefits, including the resolution of several challenges encountered in content delivery systems.

Overall, the integration of age detection and content display serves as a critical tool in optimizing content delivery systems. By utilizing this technology, companies can deliver appropriate content to the intended users while enhancing security measures. Furthermore, in the context of advertising, the system can efficiently target the appropriate user, leading to cost savings and increased engagement rates. Thus, the integration of age detection and content display presents an opportunity for companies to optimize their content delivery systems and maximize user engagement.

To achieve the integration of age detection and content display, the implementation will involve the use of cameras integrated into computer systems. In cases where a computer lacks a camera, a web camera can be installed to facilitate the process. The system will capture a facial image of the user and utilize image processing techniques to determine their age and gender. Subsequently, the content display system will display movie options that align with the user's demographic profile, ensuring the display of appropriate content.

Thus, the implementation of this technology involves the use of cameras and image processing techniques to facilitate age detection and content display. By capturing a facial image of the user, the system can accurately determine their age and gender, enabling targeted content delivery. Overall, this technology presents a significant opportunity for companies to optimize their content display systems and enhance user engagement.

2 Literature Review

In a recent study by Alex Darborg [4], a method for identifying and confirming a person's gender based on their appearance in a photograph, also known as gender prediction, was proposed. The study aimed to develop a real-time face recognition system using one-shot learning techniques, which means learning from a single or a small number of training samples. The study compared various methods to solve this problem and found that convolutional neural networks require large datasets to achieve acceptable accuracy. However, the researchers were able to achieve close to 100% accuracy by reducing the number of training instances to just one, using the concept of transfer learning. The study's findings are shown in a graph.

Ramalakshmi K [5] designed gender recognition as a technique that uses deep learning to identify a person's gender from a picture of their face. Face recognition can be affected by various factors such as pose variation, lighting, and occlusion, which can be minimized to improve the accuracy of gender prediction. The Convolutional Neural Network (CNN) was used to train the system, and faces were identified and extracted from the image to enhance accuracy. Face detection was carried out using OpenCV, which uses frontal features to identify the face, during the network training. Cropped images were used in the training dataset. The proposed method can predict gender accurately without compromising the accuracy of the system.

Sandeep Kumar [6] developed an algorithm for gender detection using face images. In this particular image processing technique called Dynamic Inflatable Histogram Equalization (BPDFHE), the brightness of the image is adjusted to make it more visually clear and appealing. The method of extracting facial features involves using the Invariable Fourier Transform (SIFT), which helps in detecting and describing unique features of an image in a scale-invariant manner. After extracting the facial features, a Support Vector Machine (SVM) classification algorithm is used to categorize and classify the features based on their characteristics.

Jinhyeok Jang [7] has developed a face attribute recognition system using Recurrent Neural Networks with visual fixation. The system calculates face features using Convolutional Neural Networks (CNN) and encodes the development of the Recurrent Neural Network function. The visual fixation is typically focused on the nose and lips or between/on the eyes. Additionally, face attribute recognition has been improved using videos. The proposed system achieved recognition rates of 89.8% for the ADIENCE dataset and 91.36% for the Multiple datasets. Computer facial recognition provides higher accuracy than human recognition, but selecting appropriate features is challenging due to redundancy and equal features.

T. Jabid and H. Kabir [8] have proposed a multi-modal identification system that combines facial and fingerprint recognition to identify individuals accurately. This system is more secure than a traditional password or numerical system as it uses physical or conduction qualities to verify identity. However, in some cases, a single character may not be accurate enough for identification, and the selected feature may not always be readable. The proposed multi-modal system is more accurate than single-feature systems and provides improved security with its two modalities.

Hugo et al. [9] have proposed an end-to-end solution for the classification challenge using the CSN network. Their solution includes an all-in-one extractor and classifier to extract soft-biometric features using a Convolutional Neural Network (CNN). The results of their study were promising as the approach demonstrated a good generalization capacity and accurately classified the three distinct qualities. Overall, their proposed method is highly promising and shows potential for effective soft-biometric feature extraction using CNN-based classification.

Kukharenko et al. [10] proposed a convolutional neural network model that allows for the automatic simultaneous extraction of multiple features in a painting. The model is based on a deep coevolutionary neural network with shared starting layers and various probabilistic outputs. Specifically, they used sex, moustache, and beard as features to identify a person's appearance. With this architecture, the neural network can identify a larger number of features without significant increases in operation time. Overall, their proposed model is promising for the efficient and accurate identification of multiple features in a painting using a convolutional neural network.

Marah Alhalabi et al. [11] proposes a system that uses deep learning algorithms to recognize the age and gender of individuals passing by an advertisement display. The system then displays targeted advertisements that are more likely to appeal to that demographic, improving the effectiveness of the advertisement while reducing waste and energy consumption associated with displaying irrelevant ads. The authors discuss the development and implementation of the proposed system, including the technology used to recognize age and gender and how it can be used to target advertisements effectively. They also explore the potential impact of the proposed system on the advertising industry, including potential benefits for advertisers and consumers. In addition, the paper examines the sustainability benefits of using targeted advertising.

3 Proposed Methodology

3.1 Proposed System Architecture

In this study, we present a methodology for predicting user age and gender based on facial recognition technology. Specifically, we utilized live video footage to capture images of the user, which were then processed using face detection algorithms to identify the facial features, which were displayed to the user using green bounding boxes. The resulting images were then analyzed by machine learning models to determine the user's predicted preferences. The Adience and Essex datasets were employed to train and validate the models. The accuracy of the classification results was assessed, and if the results exceeded a threshold of 90%, they were deemed reliable and presented to the user as recommendations. Figure 1 illustrates this methodology. The implementation of this project involved a multi-faceted approach, which incorporated several techniques, including Face Detection and Face Alignment, utilizing OpenCV for image processing, and Convolutional Neural Network (CNN) architecture for Model training. The project focused on Age group classification and Gender classification, which enabled accurate prediction of age and gender during image capture.

The present methodology involves capturing facial features using the VideoCapture() function, followed by face identification. The selected frames are subjected to pre-processing and normalization using a custom-defined function. The normalized frames are then fed into a deep learning network, which is read by means of the readmit () function following normalization.

The processed facial features are forwarded to pre-trained models, which generate prediction outcomes. These prediction outcomes are displayed visually on the screen, using green colour bounding boxes. In the initial phase, the predicted outcomes are displayed to users as videos or movies on websites that contain relevant content.

In this project, a large-scale dataset available on Kaggle was utilized. The dataset serves as a foundational resource for facial images, encompassing diverse real-world imaging scenarios, including but not limited to noise, lighting, stance, and gaze. The dataset comprises subjects categorized into eight age groups. The pre-trained models will be employed to analyze and extract information from this dataset.

3.2 System Overview

The advent of technology has greatly affected the way people access and consume media content. With the widespread use of the Internet and digital devices, content delivery systems are becoming increasingly important in ensuring that users have access to relevant and relevant content. However, while there are many content delivery systems available, most of them are generic and do not cater to the specific needs of different

AGE Based Content Display by Using Face Recognition 117

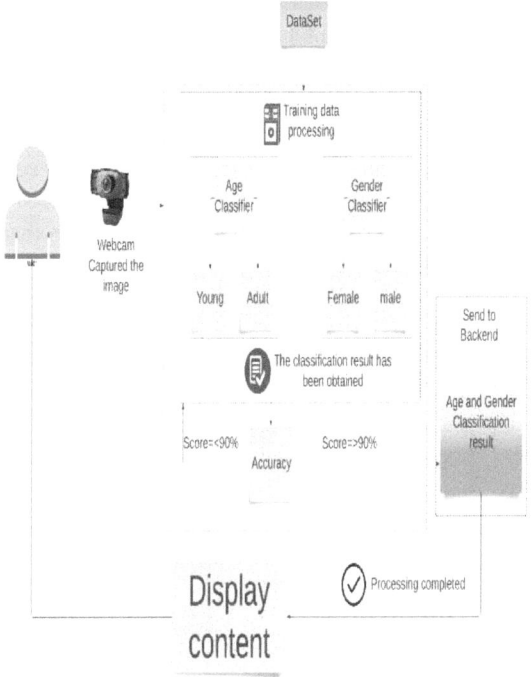

Fig. 1. Proposed System Architecture.

age groups. In response, the current study proposes a new content delivery system that is tailored to the needs of a specific age group and is sustainable in nature.

The proposed system is designed to provide appropriate content to users taking into account their age and gender. To achieve this, the system uses a built-in computer camera that continuously takes pictures of the user's face. The captured images are then processed using a convolutional neural network (CNN) that has been trained to classify and detect the age and gender of the user, and we can explain the approach utilized for retrieving the age and gender associated with the captured face from the camera, we employed the Wide Residual Network 16-8 (WRN-16-8) architecture to accurately estimate age and gender.

The selected architecture emphasizes wider networks over deeper ones, considering that training deeper models can be challenging. We chose a wide Residual Network architecture with an increased number of filters in the convolutional layers to improve its efficiency despite having fewer layers. One of the primary benefits of prioritizing width over depth is the enhanced computational efficiency it offers. Compared to the traditional ResNet [12], the Wide ResNet model used in this study has fewer layers and performs twice as fast. Specifically, we used the WRN16-8 model; the architecture includes 16 convolutional layers and a widening factor of 8, which represents the number of feature maps per layer. A 64×64 RGB image is fed as input to this network. Interestingly, even though it comprises only 16 layers, this neural network achieves comparable accuracy to a deep network with 1000 layers while being considerably quicker to train. This finding highlights the importance of residual blocks in the efficiency of a residual network with multiple layers. The output of our study is a numeric value that estimates the subject's age and gender (male or female). The resulting rating scores are saved to an online database in real time.

Then, based on the rating results obtained from the database, content from typical films is displayed on the screen. This whole process is repeated each time the user logs out of the system or website. This personalized approach ensures that users can access content that is not only relevant but also appropriate for their age and gender (Fig. 2).

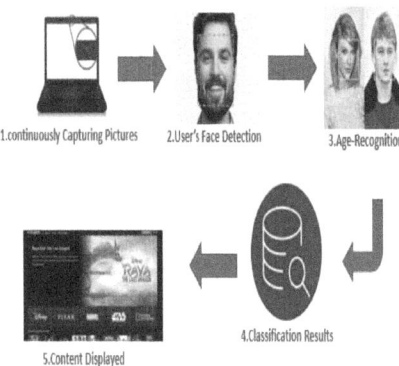

Fig. 2. Proposed System overview and work steps.

In general, the proposed system uses facial recognition technology to determine the age and gender of the user and offers appropriate movie choices based on the ratings obtained. This new approach has the potential to revolutionize the content delivery industry by providing a sustainable and personalized content delivery system that caters to the specific needs of different age groups (Fig. 3).

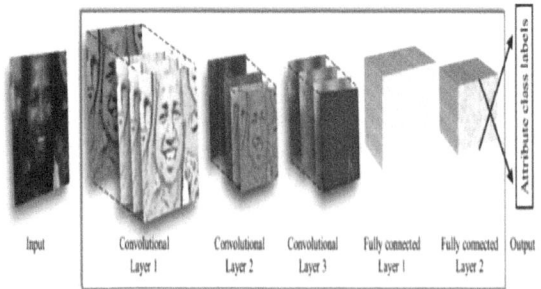

Fig. 3. An overview of CNN Architecture and its work [13].

Visual data collection is essential for many image processing applications, from object detection, recognition, and classification to facial recognition and tracking. In this context, this research project proposes to use the camera built into the laptop or to install web cameras on the computer, as shown in Fig. 4.

By using such an approach, visual data can be collected in real-time, enabling researchers to perform accurate and timely analyses. In addition, the use of a camera system eliminates the need for manual data collection, thus reducing the possibility of errors and ensuring data quality. Moreover, the proposed approach can be used in a range of research applications and has the potential to facilitate the development of more advanced and sophisticated image-processing techniques.

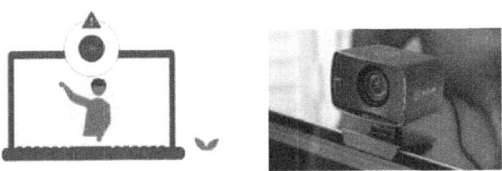

Fig. 4. The image acquisition apparatus of the system.

3.3 Algorithms

Visual gender detection employs various classification algorithms such as SVM, LDA, and AdaBoost.

3.3.1 SVM Algorithm

The support vector machine (SVM) is a machine learning technique founded on statistical learning theory and has found extensive application in pattern recognition. Unlike common classifiers such as neural networks (NN), which mainly rely on empirical risk minimization, the SVM is based on structural risk minimization, resulting in better generalization ability. In pattern recognition, the fundamental concept of SVM involves constructing a decision plane or hyperplane that can separate positive and negative patterns with the maximum margin, which entails resolving a typical Quadratic Programming problem, which can be handled using various methods. The Sequential Minimal Optimization (SMO) algorithm, which was suggested by Platt in 1998, is a widely utilized approach, but it has disadvantages such as large iteration and calculation time.

In this research paper, we introduce a novel decomposition method inspired by SMO to train SVM, which involves solving a quadratic optimization problem subject to penalty parameter C and kernel function K. The proposed method involves finding an initial feasible solution, defining a working set, and updating alpha values iteratively until a stationary point is reached. The working set selection method is used to select the most appropriate I, j values for each iteration.

It is understood that the classification vector in the transformed space adheres to the following equation:

$$\mathbf{w} = \sum_{i=1}^{n} c_i y_i \varphi(\mathbf{x}_i),$$

Here, the values are determined by resolving the optimization problem.

$$\text{maximize} f(c_1 \ldots c_n) = \sum_{i=1}^{n} c_i - \frac{1}{2} \sum_{i=1}^{n} \sum_{j=1}^{n} y_i c_i (\varphi(\mathbf{x}_i) \cdot \varphi(\mathbf{x}_j)) y_j c_j$$

$$= \sum_{i=1}^{n} c_i - \frac{1}{2} \sum_{i=1}^{n} \sum_{j=1}^{n} y_i c_i k(\mathbf{x}_i, \mathbf{x}_j) y_j c_j$$

$$\text{subject to } \sum_{i=1}^{n} c_i y_i = 0, \text{ and } 0 \leq c_i \leq \frac{1}{2n\lambda} \text{ for all } i.$$

The quadratic programming method can be utilized once more to resolve the coefficients. We can identify an index such that, which positions it on the margin's boundary in the transformed space, and then solve the equation

$$b = \mathbf{w}^T \varphi(\mathbf{x}_i) - y_i = \left[\sum_{j=1}^{n} c_j y_j \varphi(\mathbf{x}_j) \cdot \varphi(\mathbf{x}_i) \right] - y_i$$

$$= \left[\sum_{j=1}^{n} c_j y_j k(\mathbf{x}_j, \mathbf{x}_i) \right] - y_i.$$

Finally,

$$\mathbf{z} \mapsto \text{sgn}(\mathbf{w}^T \varphi(\mathbf{z}) - b) = \text{sgn}\left(\left[\sum_{i=1}^{n} c_i y_i k(\mathbf{x}_i, \mathbf{z}) \right] - b \right).$$

3.3.2 LDA Algorithm

Linear Discriminant Analysis (LDA) is a popular algorithm in machine learning and pattern recognition used for feature extraction and dimensionality reduction. It is commonly used to identify the underlying structure of a dataset by projecting it onto a lower dimensional space. LDA is a supervised learning algorithm, which means it requires labelled data to train the model. The primary goal of LDA is to find the linear combinations of features that best separate the classes in the data. These linear combinations, also known as discriminants, are mathematical expressions used to transform and project data onto a lower-dimensional space. This technique allows us to keep the maximum amount of discrimination between the data points, meaning that the differences between the data points are maintained as much as possible. By using discriminants, we can simplify complex data sets and make them easier to analyze without losing important information. It is stated that the algorithm operates by initially determining the mean and covariance matrix for each class within the dataset. Subsequently, it evaluates the within-class scatter matrix, which quantifies the dispersion of data within individual classes, as well as the between-class scatter matrix, which measures Once the optimal linear discriminants have been identified, they are used to transform the original high-dimensional data into a lower-dimensional space. This transformed data can then be used for various purposes, such as data visualization, classification, or clustering.

LDA has several advantages over other dimensionality reduction techniques, such as Principal Component Analysis (PCA). First, LDA explicitly models the class structure in the data, making it more suitable for supervised learning tasks. Second, LDA can be used to reduce the dimensionality of the data while preserving the discriminative information, which is useful for classification tasks.

$$y^* = \arg\max_y P_{Y|W}(y|\mathcal{I}_q).$$

Py\w(y\Iq), the posterior probability of class y given Iq, is computed with a combination of a probabilistic model for the joint distribution of words and classes and Bayes rule [14].

3.3.3 AdaBoost Algorithm

AdaBoost is a machine learning algorithm that is widely used to create accurate models. According to the available information, Freund and Schapire created AdaBoost in 1995. The fundamental principle of this method involves the amalgamation of several "weak" learners to generate a more precise "strong" learner. As per the provided context, individual subsets of the training data are utilized for training each weak learner. Furthermore, the weight of each subset is established by evaluating the efficacy of the previous weak learners. In the end, the predictions of all the weak learners are combined with different weights to make the final prediction. The technique aids in enhancing the precision of the model by amalgamating the strengths of numerous learners.

The procedure for mathematically calculating ε can be elucidated in the following manner:

$$\widehat{\text{MME}}_{\text{emp}}^{(j)} = \frac{\sum_{i=1}^{N} w_i I(y_i \neq h_j(x_i))}{\sum_{i=1}^{N} w_i}$$

To deconstruct this model, we can utilize the following notations:

Σ, which signifies the sum.
y_i not equal to h_j, where y_i equals 1 if the sample is misclassified, and 0 if it is classified correctly, while h_j represents our prediction.
w_i, which denotes the weight of training sample i.

In light of these notations, the formula can be interpreted as follows: "Error equals the sum of the misclassification rate, where weight for training sample i and y_i not being equal to our prediction h_j (which equals 1 if misclassified and 0 if correctly classified)."

4 Proposed System Flow Chart

In our study, Fig. 5 depicts the System architecture of the smart content display system, which involves training and assessing two Convolutional Neural Networks (CNNs), Age CNN and Gender CNN, using the ResNet50 architecture in the initial iteration.

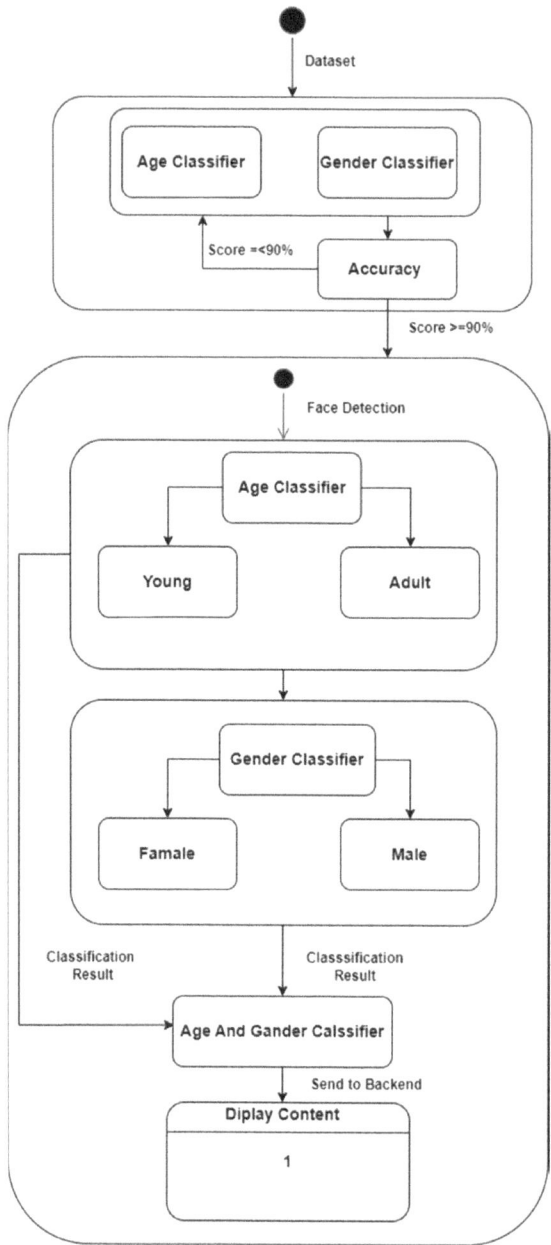

Fig. 5. System flowchart diagram.

The performance of each CNN is measured by computing their respective cross-validation scores. The threshold for both CNNs to proceed to the next phase is set at a minimum score of 90%. In the event that any of the CNNs fail to meet the threshold, the training and testing process will continue for the affected CNN until it meets the required score. Upon achieving a cross-validation score of at least 90% for both CNNs, the system proceeds to the second iteration. Here, the system detects the bystander's face, and the age and gender classify the user's age and gender, respectively. These classifications are then sent to the real-time online database, which is accessed through the website shown in Fig. 5. Based on these outcomes, the system displays videos and films that are appropriate for the user's age and gender. This second iteration operates continuously every time the user attempts to exit the system or the front-end website.

Our approach employs the ResNet50 architecture in training and assessing the CNNs, which is a widely used and effective user. Our approach employs the ResNet50 architecture in training and assessing the CNNs, which is a widely used and effective approach in computer vision tasks. The use of real-time online databases and website access ensures that the system can quickly adapt to changes in user data and preferences. By implementing this system, we aim to provide a more personalized and efficient content viewing experience for users.

5 Results

The aim of this research project is to create a facial recognition system that can effectively identify the age and gender of a person with high accuracy. By successfully developing such a system, it could have numerous practical applications in fields such as law enforcement, security, and marketing research, among others. The proposed system is expected to find diverse applications in several fields, such as marketing, advertising, and safety. The main objectives of this research project are to determine the age and gender of the target group for the advertisement, or that uses the system and to present suitable content for this group.

To achieve these objectives, the proposed system will be built using machine learning techniques, utilizing appropriate algorithms and tools such as Flask, OpenCV, and ResNet. These techniques will enable the system to process large amounts of visual data and identify specific facial features and patterns that can be used to accurately determine an individual's age and gender.

The utilization of machine learning algorithms provides the system with the ability to learn and adapt to different facial features and expressions, ensuring its accuracy and effectiveness in determining age and gender. Additionally, the proposed system's deployment on suitable platforms such as Flask and OpenCV ensures its scalability, portability, and compatibility with a wide range of devices (Figs. 6 and 7).

In the beginning, the code is processed, and the model is executed on the image captured from the live video by reading the face and highlighting it using the face box. And after processing the face and applying the trained data, the age and gender appear as a result on the screen (Fig. 8).

And then, the target group of the advertisement or those who want to browse the Internet is determined to ensure the safety of browsing for the user and to display the appropriate content.

One of the security features of the age-and-gender system is the ability to distinguish users over 18 years of age, allowing unrestricted access to movie platforms such as

Realtime_demo.py

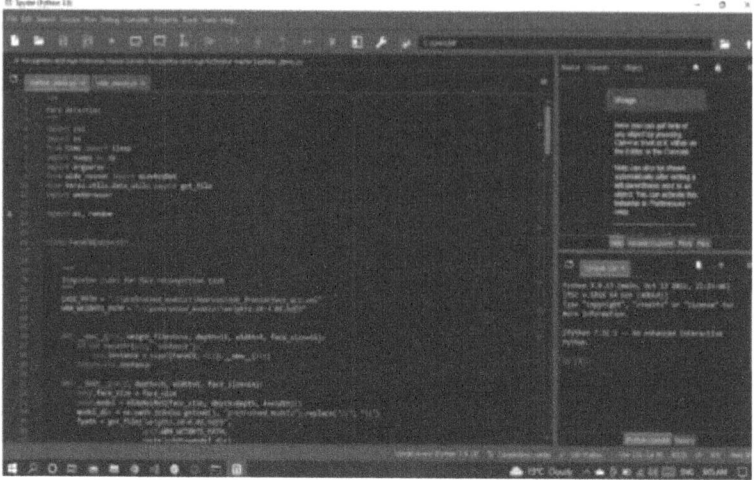

Fig. 6. Realtime_demo.py Code.

Wide_resnet.py

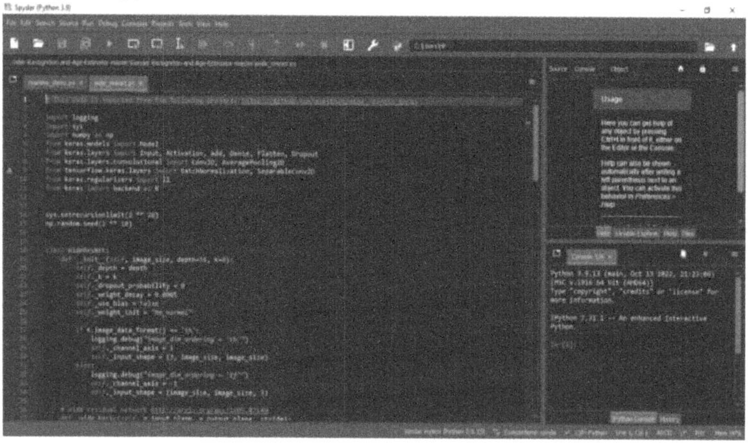

Fig. 7. Wide_resnet.py Code.

Netflix. This feature ensures that users are not restricted from accessing desired content while also preventing minors from accessing inappropriate content (Fig. 9).

But if the user is −18, then special web pages are displayed that restrict him from accessing films that do not suit his age and display appropriate content instead.

The integration of the age and gender determination system into the marketing systems for individuals and companies is expected to lead to important results. Specifically, the accuracy in target group identification is expected to improve by adding different trained images and data. Moreover, the system will enhance accessibility and security, ensuring that the correct advertisement is served to the target audience. If the system

AGE Based Content Display by Using Face Recognition   125

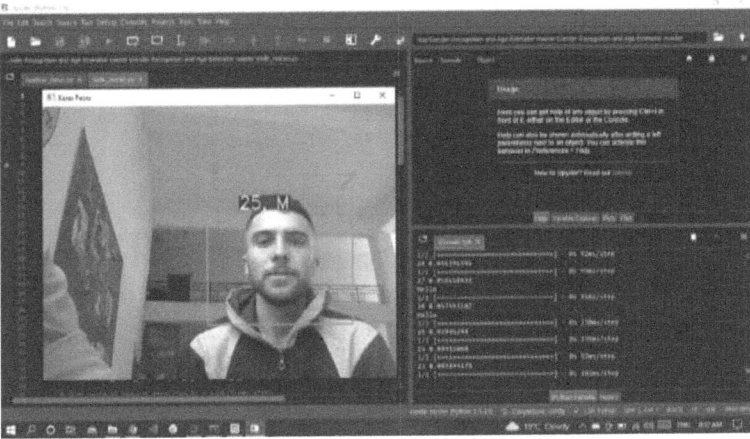

Fig. 8. The facial detection using the trained model and detect the users age in Realtime.

Fig. 9. Age-Appropriate Content Website for Users Aged 18.

is intended for security and restriction purposes, browsing restrictions will apply if the results are within the restricted category. This ensures safe and efficient surfing.

The application of the age and gender determination system will reduce the possibility of errors in reaching the target groups of the advertisement or restrict groups from accessing inappropriate content. In addition, the system efficiency is expected to be greatly improved because the automated system can operate much faster than the human operator, thus improving the efficiency rate and overall quality. Our proposed approach for classifying age and gender uses Convolutional Neural Networks (CNNs) on two datasets: the Essex dataset and the Adience benchmark, which have been utilized in this study. The primary objective is to precisely predict the gender and age category of an individual. The methodology employs CNNs for both gender and age classification.

Fig. 10. Age-Appropriate Content Website for Users Aged −18 after the recognizing of the age by the system.

The system is designed for a multiclass classification task involving age and gender classification. To achieve this, the model is trained using a loss function that is specifically designed for classification-based training targets. The input image undergoes pre-processing prior to training and is then flattened into a one-dimensional vector feature that accounts for both age and gender. These features are then input into a Convolutional Neural Network (CNN) for classification. The CNN outputs are connected to three fully connected layers, which are optimized using the SoftMax function for the age categories (0–10, 10–25, 25–30, 30–35, 35–40, 40–45, and +45 years old) and the sigmoid function for gender categories (male and female) (Tables 1, 2 and Fig. 12).

Table 1. Essex dataset.

Age classes	0–10	10–25	25–30	30–35	35-40	40–45	+45
Acc (%)	85.5	91.2	95.4	97.1	99.5	99.1	97.7

Table 2. Adience benchmark.

Age classes	0-10	10–25	25–30	30–35	35-40	40–45	+45
Acc (%)	81.5	85.6	91.3	90.7	91.1	96.7	93.4

The proposed network in this study demonstrates significant improvements in accuracy for age and gender classification, as seen in Fig. 10, 11 and Table 3. However, it requires a long training time, with a minimum of 36 h needed for training with an 80,000

Table 3. Gender and age classification: Comparing Our Gender Classification Performance with State-of-the-Art Approaches.

Method	Dataset	Accuracy Gender %	Accuracy Age %
A.Ekmekji [15]	Essex	N/A	N/A
	Adience benchmark	84.7	51.2
A.Olatunbosun et al. [16]	Essex	N/A	N/A
	Adience benchmark	97.1	94
S.M. Osman et al. [17]	Essex	92.95	93.3
	Adience benchmark	N/A	N/A
Benkaddour, Mohammed [18]	Essex	99.10	95.41
	Adience benchmark	96.3	92.5
Our approach	Essex	98.98	94.95
	Adience benchmark	97.2	89.89

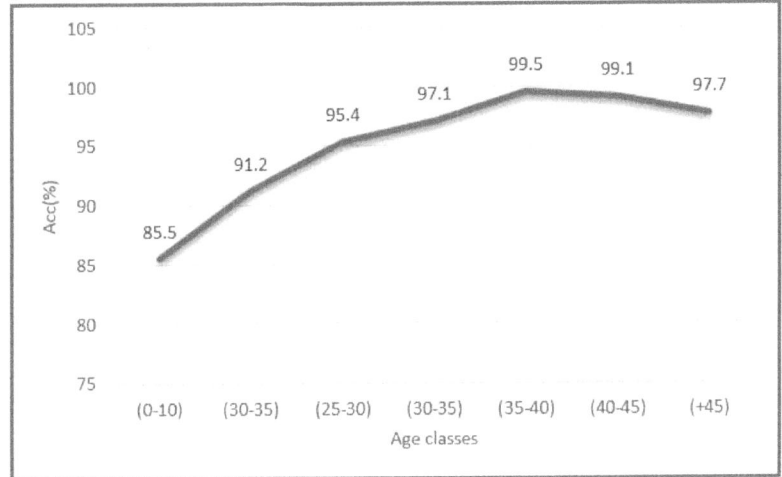

Fig. 11. Age accuracy rate (Essex).

ERA. The proposed approach performs well in adult age groups but struggles with very young subjects, which is expected since age estimation is more challenging for young children. Gender errors are also common in images of babies or young children without clear gender features.

Table 4 compares the proposed model's accuracy to recent methods of age and sex classification, showing improvements in accuracy compared to previous work.

Despite encountering challenging viewing conditions in the images from the Essex and Adience benchmark datasets, the models presented in this study were able to attain an average accuracy rate, albeit with some misclassifications. In terms of age and gender

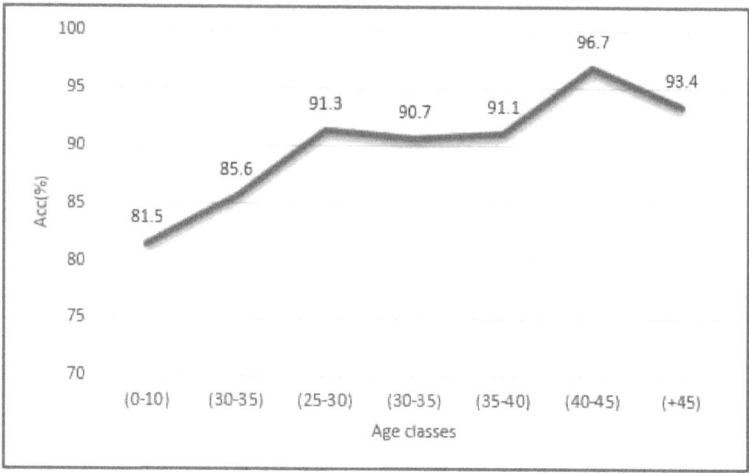

Fig. 12. Age accuracy rate (Adience benchmark).

classification, our method surpassed the latest techniques and produced the best results for both the Essex face dataset and the Adience benchmark.

Overall, the project's objective of delivering age-appropriate content with accuracy and a low error rate has been successfully achieved.

6 Conclusion

Technological advances have led to the development of more efficient and sophisticated facial recognition systems that detect age and gender through trained data and machine learning methods. These systems have the potential to be used in various fields, such as marketing and security, by using machine learning algorithms to classify target groups. Integration of age and gender determination systems with corporate advertising and marketing systems, or systems for personal use, can yield significant benefits in terms of improving quality, ensuring safety, and reducing errors. Automated systems can operate faster and guarantee more features and benefits compared to manual operators, thus increasing work efficiency, saving time and ensuring effective results. Implementing such systems can save cost and time, as they operate continuously without the need for manual labour and can carry out repetitive tasks, allowing employees to focus on higher-level tasks. Moreover, the system can ensure safe browsing environments for individuals, thus helping to achieve the desired goals and enhance Internet surfing. It can be concluded that integrating face recognition systems to detect gender and age into personal or marketing systems has many benefits, including improved efficiency, accuracy, browsing quality, and security. With advances in technology and machine learning, these systems are expected to be more sophisticated and widely adopted in many fields, leading to increased benefits.

Acknowledgement. It is a pleasure for us to acknowledge the assistance and support of work under the guidance for this research Dr.K.S. Arvind, program head Dr. Chandramma R and the faculty of Engineering and Technology, Jain (Deemed-to-be) University.

References

1. Facta, M.: Institute of Electrical and Electronics Engineers. Indonesia Section, Universitas Diponegoro. Department of Electrical Engineering, and Institute of Electrical and Electronics Engineers, Proceedings, 2016 3rd International Conference on Information Technology, Computer, and Electrical Engineering (ICITACEE 2016), Semarang, Indonesia, October 19–21 (2016)
2. Hakhoe, H.T'.: IEEE Communications Society, Denshi JōhōTsūshin Gakkai (Japan). Tsūshin Sosaieti, and Institute of Electrical and Electronics Engineers, IEEE ICAIIC 2019: the 1stInternational Conference on Artificial Intelligence in Information and Communication, Okinawa, Japan, 11–13 February 2019 (2019)
3. Khryashchev, A.: Applied Imagery Pattern Recognition Workshop (AIPR). IEEE (2014)
4. Madasu, V.K., et al.: Synergy of Automatic Machine Recognition and Human Validation for Match Accuracy Improvement of Biometric Identification Systems Teddy Ko, Raytheon Session: Performance Analysis and Modeling Session Chair: Robert Mericsko 10:50 The Use of Operator Feedback to Reduce False Alarms in Hyperspectral Systems Objective Performance Evaluation of a Moving Object Super-Resolution System
5. Jemima Jebaseeli, T.: Prediction of gender from facial image using deep learning techniques. J. Mech. Continua Math. Sci. **15**(2) (2020). https://doi.org/10.26782/jmcms.2020.02.00010
6. Kumar, S., Singh, S., Kumar, J.: Gender classification using machine learning with multi-feature method (2019)
7. Jang, J., Cho, H., Kim, J., Lee, J., Yang, S.: Facial attribute recognition by recurrent learning with visual fixation. IEEE Trans. Cybern. **49**(2), 616–625 (2019). https://doi.org/10.1109/TCYB.2017.2782661
8. Jabid, T., Kabir, M.H., Chae, O.: Gender classification using local directional pattern (LDP). In: Proceedings - International Conference on Pattern Recognition, pp. 2162–2165 (2010). https://doi.org/10.1109/ICPR.2010.373
9. Shobeirinejad, A., Gao, Y.: Gender classification using interlaced derivative patterns. In: Proceedings – International Conference on Pattern Recognition, pp. 1509–1512 (2010). https://doi.org/10.1109/ICPR.2010.1118
10. Mozaffari, S., Behravan, H., Akbari, R.: Gender classification using single frontal image per person: combination of appearance and geometric based features. In: Proceedings - International Conference on Pattern Recognition, pp. 1192–1195 (2010). https://doi.org/10.1109/ICPR.2010.297
11. Alhalabi, M., Hussein, N., Khan, E., Habash, O., Yousaf, J., Ghazal, M.: Sustainable smart advertisement display using deep age and gender recognition. In: 2021 International Conference on Decision Aid Sciences and Application, DASA 2021, pp. 33–37 (2021). https://doi.org/10.1109/DASA53625.2021.9682398
12. Zagoruyko, S., Komodakis, N.: Wide residual networks, May 2016. http://arxiv.org/abs/1605.07146
13. Challa, S.T., Jindam, S., Reddy, R.R., Uthej, K.: Age and gender prediction using face recognition. Int. J. Eng. Adv. Technol. **11**(2), 48–51 (2021). https://doi.org/10.35940/ijeat.B3275.1211221

14. Rasiwasia, N., Vasconcelos, N.: Latent dirichlet allocation models for image classification. IEEE Trans. Pattern Anal. Mach. Intell. **35**(11), 2665–2679 (2013). https://doi.org/10.1109/TPAMI.2013.69
15. Ekmekji, A.: Convolutional neural networks for age and gender classification. Technical report, Stanford University (2016)
16. Anand, A., Labati, R.D., Genovese, A., Munoz, E., Piuri, V., Scotti, F.: Age estimation based on face images and pre-trained convolutional neural networks. In: Proceedings of the IEEE Symposium Series on Computational Intelligence (SSCI), Honolulu, HI, USA, November 2017, pp. 1–7 (2017)
17. Osman, S.M., Noor, N., Viriri, S.: Component-based gender identification using local binary patterns. In: Nguyen, N., Chbeir, R., Exposito, E., Aniorté, P., Trawiński, B. (eds.) ICCCI 2019. LNCS, vol. 11683, pp. 307–315. Springer, Cham (2019). https://doi.org/10.1007/978-3-030-28377-3_25
18. Benkaddour, M.: CNN based features extraction for age estimation and gender classification. Informatica **45** (2021). https://doi.org/10.31449/inf.v45i5.3262

IoT Enabled Heart Disease Accuracy Prediction of Healthcare Dataset Using Deep Belief Network

Rahama Salman[1(✉)] and Subodhini Gupta[2]

[1] Department of Computer Science, Sam Global University, Bhopal, MP, India
guddu.rahama@gmail.com
[2] Department of Computer Application, School of Information Technology, Sam Global University, Bhopal, MP, India

Abstract. Fog is a data management and analytics service. This paper derives the most effective new approach for providing IoT-enabled services in healthcare applications using Fog Computing. In this study, data are collected from Google Scholar, Science Director, and MEDLINE databases. IoT-based Fog Computing techniques are proposed to provide quality services to users. An optimal resource provisioning method for boundary discovery, service level agreements, and administration services for an IoT client is proposed. The DeepQ residual information processing technique is applied to cloud data center connectivity, and the computing paradigm technique is to find the reference depth of fog levels. The proposed optimal resource provisioning algorithm studies the dataset and the TensorFlow tool is used to simulate the environment. The Deep Belief network is generated based on the above input using a $512 \times 512 \times 3$-layer system and 3000 trained data, 1000 test data are taken for simulation. Each dataset simulation is recorded using supervised and unsupervised learning techniques. Based on the above results, IoT enables Fog Computing's data management and analytics systems to provide 95% accuracy, and compared to existing computing methods, our proposed systems show better performance in terms of security and convenience.

Keywords: Medical Dataset · Prediction · Accuracy · Deep Learning · TensorFlow

1 Introduction

Towards Industry 4.0 such as Internet of Things, Cloud Computing, Big Data and Deep Learning. The huge volume of data or dataset are processed and accessed. So, we need to access the data more quick and efficient manner [1]. The concepts of Fog computing will play the important role to access the data from cloud and do the big data analytics operations. The cloud services can be accessed by different computing devices and each device are represented as edges. Saleforce.com [2] is started fog computing trends with combination of cloud, IoT and Data analytics process for healthcare applications [3].

Current situation healthcare applications are growing tremendously. Lot of online based consulting and medicine facility are coming and most of the persons are using the online services effectively [4]. In Google survey mentioned that past two years more the 3 billion of people are used online healthcare services [5]. The home-based services are adopted and supported by government also introduced door by door vaccination process with automated services. It is the smart model with IoT enabled services to monitor the patients and periodically check the status. All the monitoring process are documented and recorded [6]. Manzoor and Chigao et al., provided IoT [7], Cloud and Fog computing are current key players and these three technologies leads to access the remote data services. Healthcare industries are coming to smart based automation services. Edge computing is the combination for three building blocks to handling dataset and provide efficient [8, 9]. Gigoer et al., Cloud and Fog Computing are major computing paradigm to proceed home based healthcare monitoring applications. Edge Computing is assisting the healthcare industries to communication and making decisions [10].

Manikandan et al., the cloud data can be transmitted to healthcare repository and apply the data analytics process so we can give exact process to the user. This approach will give more impact on device accessing level moreover opinion based survey [11]. PubMed dataset is collected from Google Scholar and which gives result of MEDLINE [12] information. In recent times, IoT and Cloud services data can be managed by

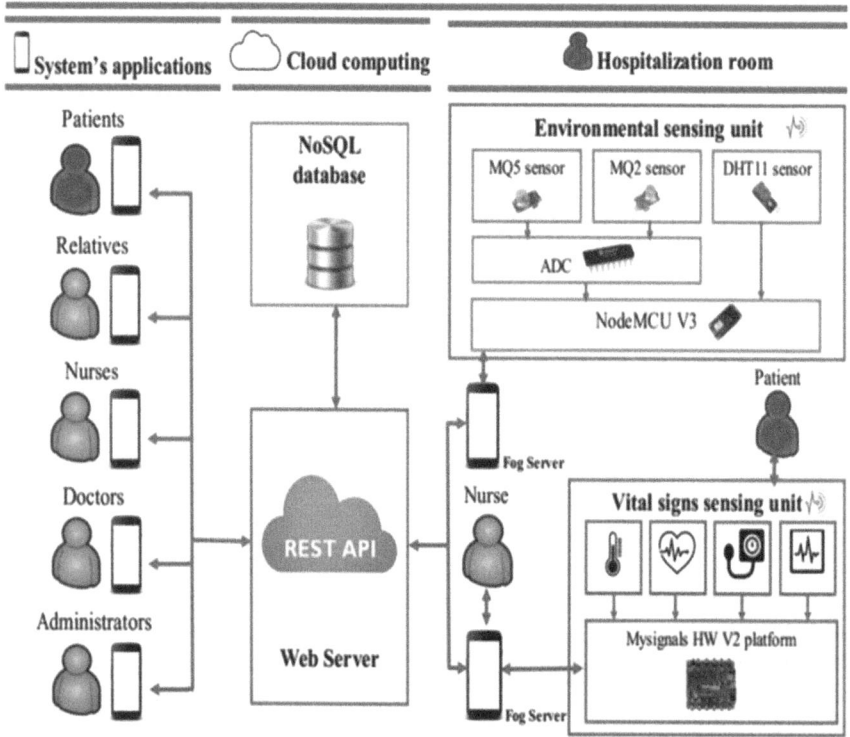

Fig. 1. Data Processing System –Proposed Management Model

Boolean operation. This method enables to handle multiple research-based services to handles the operations. The data is collected from edge devices and sensors are sensing the input and access the resources [13]. From below Fig. 1, from the patient side body sensor one edge to collect the raw data and stored in server. User applications are running and create the optimal resource provisioning decision support procedure for data analytics. The data can be process and responding based on timing. The fog nodes are created by using following procedure.

Fog node can analyze the raw data which is received from edge device and sensitive data are marked. Data management process is taken such as data modelling, association, clustering and normalization operation. The least sensitive or redundant data removed and analyze fog node contents. IoT enabled service so the data can be collected from multiple sources and fogging process is applied in all edges, nodes and cloud platforms.

2 Proposed Methodology-Deep Belief Network

An optimal method of providing resources for processing fog data and measuring the accuracy coefficient is proposed. In data management, we need to take care of data throughput, minimum delay value, real-time interaction ratio, distribution values, mobility index and business continuity. The data should be analyzed on the server and the best performance value should be chosen as the index. The process diagram below shows that the data management process is performed using an optimal resource provisioning method.

The proposed system focuses on the integration of IoT and fog computing to process the dataset and implement the sensing system. From this optimal resource provisioning method, there is a node module, a detection system, a power protection system, and a detection module. The efficiency calculation algorithm is proposed below (Fig. 2).

The optimal resource provisioning method has fuzzy Markov predictions that select user demand parameters such as cloud service provider representation, fog server value, and services X is the service and the suppliers are P (p1, p2 ... pn). The data sampled from the cloud service provider and the Markov prediction are represented as

$$F\infty \{SP \leq X(P) / SP(Y) * P(Y)\} \tag{1}$$

The ranking mechanism is applied to the data management process, which has decomposition, aggregation and prioritization. Membership is calculated as

$$Ax = \{x \in P;\ A(x) \geq SP\} \text{ for any } N \in [0, 1] \tag{2}$$

$$Ay = \{y \in P;\ A(y) > SP \text{ for any } N \in [0, 1] \tag{3}$$

The entry amount is calculated as

$$HWt = \frac{\sum_{i=1}^{N} D_i(x)}{N} \tag{4}$$

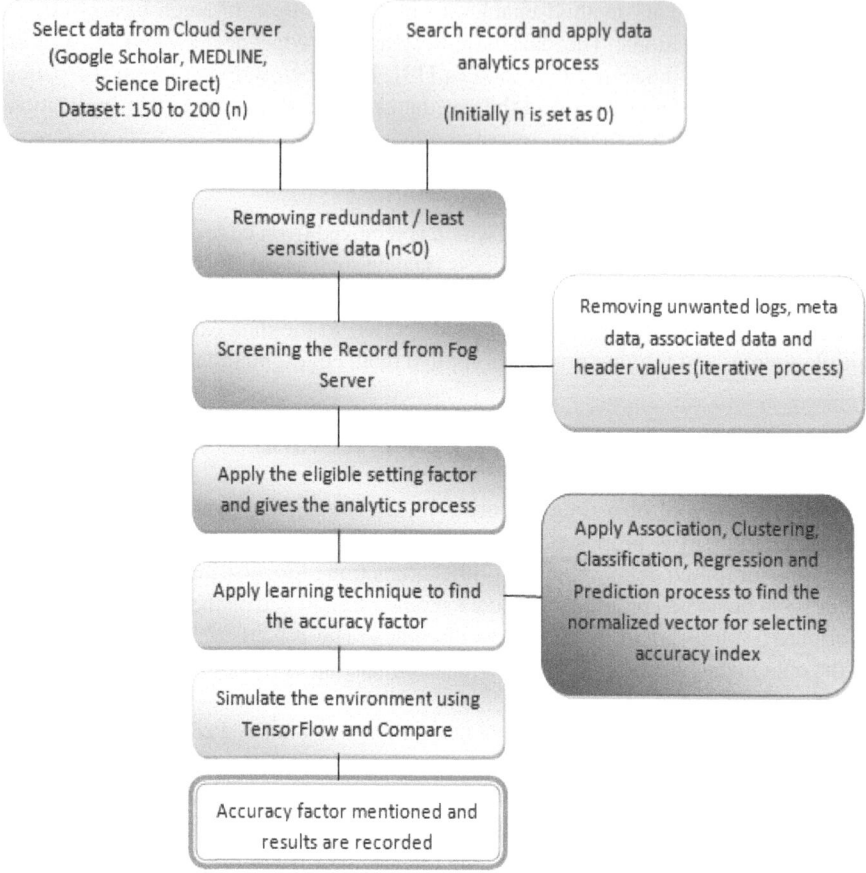

Fig. 2. Proposed Optimal Resource Provision System

Whereas N is represented as number of resources taken for evaluation and A is accuracy factor measure calculated from ranking model. The accuracy is calculated using the membership value and the amount of entropy. This is the average value measured over the transmitted information. The number of logs, events, probability values and wait are recorded.

3 Simulations – Tensor Flow

In this section, the dataset classified and record the logs(P), events (Q). Based on the results DeepQ network is generated by using below representations,

Input: (P, Q) ≡ P: Training and test data set $512 \times 512 \times 3$ layers
Output: Data management – Accuracy ratio of optimal resource provisioning
Step 1: Each value is recorded based on the fog data and the DeepQ process is applied
Step 2: Calculate DeepQ(P, Q) = $(1/N) \sum_{}(i=0) \wedge (p-1) \, G \, F(i) - \sum(i-0) \wedge (p-1) \, X[V(i) - V(minx)]]$

Step 3: Find The accuracy factor is measured by DeepQa = $\sum(i = 0) \wedge (p - 1)$ G/((V(i) - S(i)/N))
Step 4: Write down the stored value and repeat the process until n < 0

The above model provides a fog-based precision health monitoring system, and the experiments are performed using Google TensorFlow. Here, sensor layer, fog server layer and cloud service layer are taken for performance evaluation and analysis (Fig. 3).

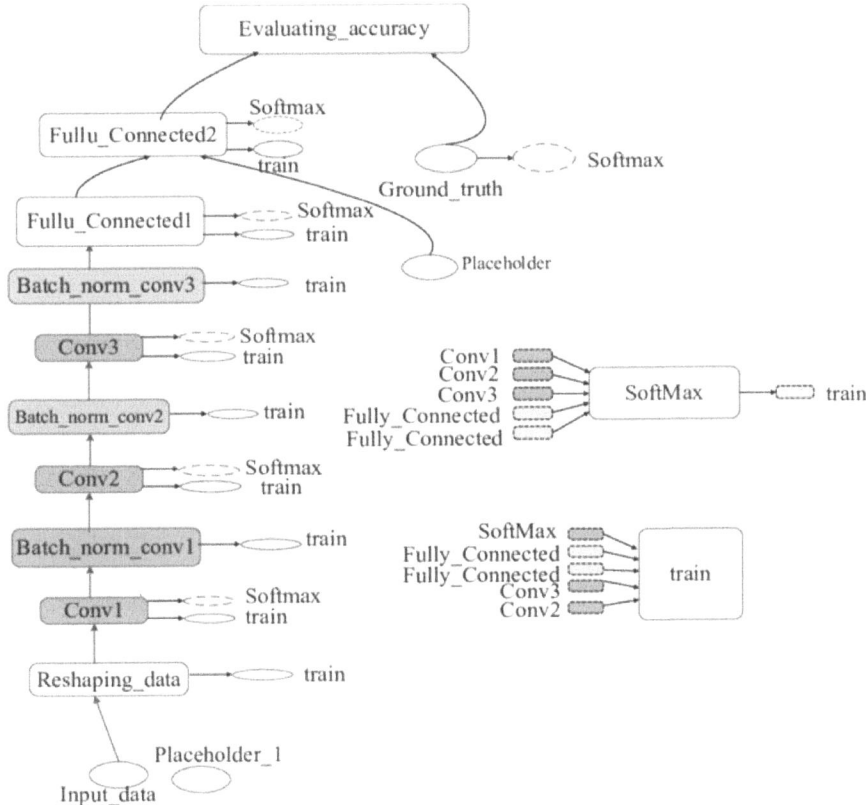

Fig. 3. TensorFlow Simulation of Deep Belief Network Optimization

The above is a deep convolutional network created using Google TensorFlow input conditions. This method included smart system recommendations and fog-based results. It collects records and applies layer-by-layer strategies. The table below shows that the data analytics values obtained from the fog server and accuracy dependencies with node 512 × 512 × 3 (Table 1).

From the above table, compare with machine learning and deep learning algorithm with different tools. Our proposed system is simulated using TensorFlow and has the best accuracy rate of 95% on 3000 trained data and 1000 test data with values of 512 × 512 × 3 hidden layer nodes 8, 16, 32, 64. Optimum management of resource provisioning data

Table 1. Accuracy result of proposed Healthcare Dataset using TensorFlow

Iterations	Hidden Values	Dimension	Accuracy	Average	Precision
1	8,16,32	500,250,100	98	95	0.18
2	8,16,32	500,250,100	97	94	0.19
3	8,16,32	500,250,100	96	95	0.23
4	8,16,32	500,250,100	97	96	0.24
5	8,16,32	500,250,100	98	94	0.27
6	8,16,32	500,250,100	96	95	0.31
7	8,16,32	500,250,100	97	96	0.34
8	8,16,32	500,250,100	98	95	0.38
9	8,16,32	500,250,100	96	96	0.42
10	8,16,32	500,250,100	97	95	0.52

provides a precision ratio with respect to accuracy and memory ratio with TensorFlow. The Table 2 shows that comparison of proposed method with existing algorithms which is shown in below.

Table 2. Comparison of Existing method with proposed accuracy results

Methodology	Dimension	Accuracy	Precision
Support Vector Machine	500,250,100	68%	3.14
MLearn	500,250,100	71%	2.27
DelMedi	500,250,100	78%	1.32
Proposed Method	500,250,100	95%	0.32

4 Conclusion

The DeepQ analysis method is used to evaluate the data set and measure the accuracy ratio using the data analysis model. This approach provided a better accuracy rate and compared to existing methods. An IoT-enabled fog server is used to identify the user and record values in the cloud. The training and testing data are simulated using TensorFlow with a 512 × 512 × 3-layer deep belief network and achieved 95% accuracy. In the future, fog computing will be extended for intelligent decision-making approaches.

References

1. Manikandan, S., Chinnadurai, M.: Virtualized load balancer for hybrid cloud using genetic algorithm. Intell. Autom. Soft Comput. **32**(3), 1459–1466 (2022)
2. Yeole, A.S., Kalbande, D.R.: Use of Internet of Things (IoT) in healthcare: a survey. In: Proceedings of the ACM Symposium on Women in Research (WIR 2016), Indore, India, March, pp. 71–76. Association for Computing Machinery, New York (2016)
3. Tahir, S.; Bakhsh, S.T., Abulkhair, M., Alassafi, M.O.: An energy-efficient fog-to-cloud Internet of Medical Things architecture. Int. J. Distrib. Sens. Netw. (2019)
4. Manikandan, S., Chinnadurai, M.: Intelligent and deep learning approach OT measure e-learning content in online distance education. Online J. Distance Educ. e-Learn. **7**(3) (2019)
5. Sridharan, M., Arulanandam, D.C.R., Chinnasamy, R.K., Thimmanna, S., Dhandapani, S.: Recognition of font and Tamil letter in images using deep learning. Appl. Comput. Sci. **17**(2), 90–99 (2021)
6. Fortino, G., Savaglio, C., Spezzano, G., Zhou, M.: Internet of Things as system of systems: a review of methodologies, frameworks platforms, and tools. IEEE Trans. Syst. Man Cybern. Syst. **51**, 223–236 (2021)
7. Manikandan, S., Chinnadurai, M., Maria Manuel Vianny, D., Sivabalaselvamani, D.: Real time traffic flow prediction and intelligent traffic control from remote location for large-scale heterogeneous networking using TensorFlow. Int. J. Future Gener. Commun. Netw. **13**(1), 1006–1012 (2020)
8. Chettri, S., Debnath, D., Devi, P.: Leveraging digital tools and technologies to alleviate COVID-19 pandemic. SSRN Electron. J. (2020)
9. Al-khafajiy, M., et al.: Remote health monitoring of elderly through wearable sensors. Multimedia Tools Appl. **78**(17), 24681–24706 (2019). https://doi.org/10.1007/s11042-018-7134-7
10. Rajabion, L., Shaltooki, A.A., Taghikhah, M., Ghasemi, A., Badfar, A.: Healthcare big data processing mechanisms: the role of cloud computing. Int. J. Inf. Manag. **49**, 271–289 (2019)
11. De MoraisBarrocaFilho, I., Aquino, G., Malaquias, R.S., Girão, G., Melo, S.R.: An IoT-based healthcare platform for patients in ICU beds during the COVID-19 outbreak. IEEE Access **9**, 27262–27277 (2021)
12. Silva, C.A., Aquino, G.S., Melo, S.R.M., Egídio, D.J.B.: A fog computing-based architecture for medical records management. Wirel. Commun. Mob. Comput., 1–16 (2019)
13. Manikandan, S., Dhanalakshmi, P., Priya, S., Mary OdilyaTeena, A.: Intelligent and deep learning collaborative method for e-learning educational platform using TensorFlow. Turk. J. Comput. Math. Educ. **12**(10), 2669–2676 (2021). E-ISSN: 1309-4653

Handwriting Analysis for Bank Cheque Verification Using EfficientNet

Jaydeep Ranpariya, Roshan Saravanan, Alvin James, Pranav Abraham, and S. Ramesh(✉)

Department of Computer Science and Engineering, Jain (Deemed-to-be University),
Ramanagara District, Bengaluru 562112, Karnataka, India
`{19BTRCR010,19BTRCR012,19BTRCR052,19BTRCR054, ramesh.s}@jainuniversity.ac.in`

Abstract. Cheque analysis and verification is a challenging problem in financial technology that still has yet to have solutions that are competent and efficient, and most importantly, accurate at detecting fraudulent cheques. This paper proposes the approach with the help of a convolutional neural network for handwriting analysis for bank cheque verification. The development of an automated system that can verify the authenticity of bank cheques by comparing the handwriting on the cheque with the handwriting of the account holder on record as well as handwriting samples within the cheque itself. We extract handwriting features from the cheque and compare them with the handwriting samples from various parts of the cheque as well as those within the bank database to ascertain feature similarity and report anomalous samples. This method provides a promising approach for the automated verification of bank cheques enabling an improvement in efficiency while enhancing the security and reliability of the cheque verification pipeline employed by various financial institutions.

Keywords: YOLO · Text Identification · Optical Character Recognition · Patch-scanning · Convolutional Neural Networks · Text-independent · Handwriting Analysis

1 Introduction

Handwritten bank cheques remain a common means of transaction globally, albeit they are currently being replaced with more modern methods like digital transfers, bank-to-bank transfers, debit/credit cards, etc. Bank cheques are still a relevant service offered by thousands of financial institutions over the globe. Cheques are used by over 70% of the senior citizen population across the world and there are reported frauds upwards of ₹1.58 billion annually. Such extensive usage and fraud rates make it crucial to ensure their security and authenticity. One of the most common forms of cheque fraud by a bad actor is the manipulation of information handwritten onto a cheque by the payer. A key element to detecting and verifying the authenticity of a cheque–often ignored in automated methods–is handwriting analysis and comparison. Handwriting can be seen as a form of biometric for an individual and yet, determining handwriting similarity can be deemed challenging due to the vast variations in writing styles, as well as the wide diversity of potential for handwriting forgeries. Identification of an individual has been applied in forensics and historic documentation studies, but it proves to be a formidable task to tackle that takes a lot of domain expertise and time to make conclusions.

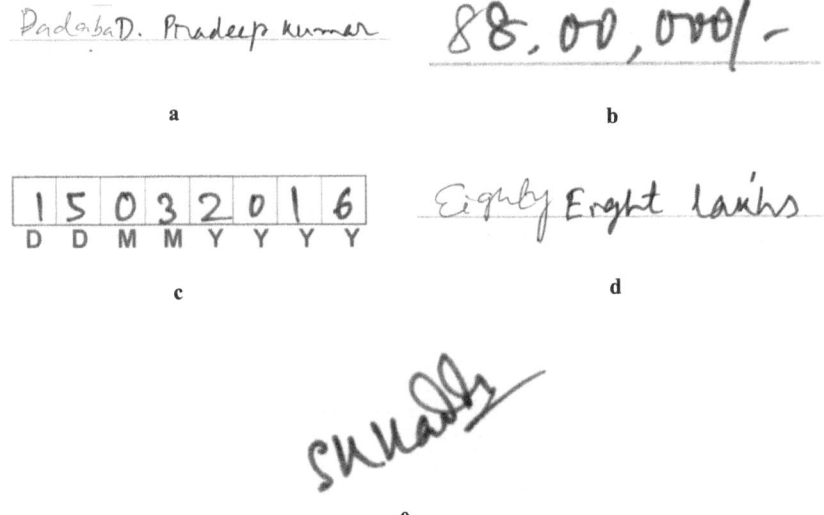

Fig. 1. a. Handwritten Name of Payee **b.** Handwritten Amount (in words) **c.** Handwritten Date **d.** Handwritten Amount (in digits) **e.** Signature of Payee

There are two primary types of handwriting analysis: offline identification [3], and online identification [7]. Off-line identification deals with tabular temporal data that contains features pertaining to the pressure/angles/pen tip positions of writing strokes. Off-line identification instead takes images of handwriting as a whole and uses that for performing analysis and identification. Within offline identification [3], there is further categorisation into text-independent and text-dependent classification. Text-dependent classifications rely heavily on the two patches of handwriting being compared to containing the same text. This method is less computation intensive and yields more accurate results while the more challenging form of classification is text-independent where the writing style of the ink strokes needs to be extracted and compared between texts that contain different information to produce a similarity score.

We put forward a text-independent model for sequentially identifying and extracting handwritten data from a bank cheque and then performing handwriting analysis and verification using deep CNNs to draw out powerful representations of the local handwritten patches and optimise a consecutively levelled structure for the writer identification task. The developed structure collates the patches of handwriting against bank-stored records as well as other patches of handwriting samples procedurally extracted from the bank cheque itself. Deep convolutional neural networks have proven their efficacy in multiple image-based problem domains, providing best-in-class results for image classification, facial recognition, and object detection. Additionally, we present data augmentation techniques [5] to improve the classification accuracy of our model and a patch scanning approach [1] to deal with text images of varying sizes.

The performance and effectiveness of our model are evaluated on the IDRBT dataset [4] and its performance is compared against prior methods [1–3]. Figure 1 shows various patches of handwritten information extracted from a bank cheque through the initial image processing pipeline.

The experimental results showed that the proposed approach of using patch scanning, data augmentation, and feature extraction using EfficientNet achieved high accuracy in handwriting analysis for bank cheque verification. The model achieved an overall accuracy of 99.4% accuracy on the test set, indicating that it can accurately distinguish between writers for bank cheque verification.

2 Literature Review

Multi-stream Deep Convolutional Neural Network (CNN) [1] approach uses multiple image representations of the handwritten text to capture different aspects of the handwriting style. The network consists of three streams: a stroke-based stream that processes the dynamic information of the handwriting, a texture-based stream that captures the static information of the handwriting, and a direction-based stream that extracts the directional features of the handwriting. The three streams are combined to produce a final prediction of the writer's identity.

There are basically five steps before training in which authors preprocess the images before training the model. This includes the creation of multi streams followed by patch scanning [1], tuning kernel size and neural number and finding out common shared features or patterns which are found across training datasets. It is experimentally demonstrated that accounting for spatial relationships between various image patches tends to benefit writer identification.

The multi-stream deep CNN [1] performs robustly in text-dependent writer identification, managing reasonable accuracy in identifying different variations of handwriting. Conversely, the drawback of the system is that it does not address the offline text-independent writer verification function. This limitation may restrict the system's application in scenarios where only a limited amount of text samples are available for a given writer. Additionally, the system does not currently support multi-task learning of identification and verification, which may limit its ability to adapt to different use cases and scenarios.

Siamese Time Delay Neural Networks have been used in the domain of signature verification [2] to create a system based on a Signature Capture Device and to use eighty bytes or less for convenient signature feature storage. A deep learning convolutional model called a "Siamese" Neural Network [2] is used, consisting of two similar sub-networks merged at their final outputs, which draws out features from 2 signatures and calculates the distance between the 2 feature vectors during training.

Before inputting the signature into the neural network, several preprocessing steps are taken. The signature is resampled to have a length of 200 points using linear interpolation, and any remaining features are calculated as input to the network. All input values are then scaled so that the majority falls within a certain range. The Siamese neural network used in this algorithm has two input neurons to compare two images and one output whose value gives the corresponding similarity score. Two other sub-networks based on Time Delay Neural Networks [2] are used on each input image to extract features, and the angular cosine between two feature vectors is computed, which represents the distance. The network is trained using a modified version of backpropagation, with all weights able to be learnt, but the two sub-networks are bounded to have identical parameters. The best performance was obtained with Network [2] 4, with 95% of genuine signatures detected when the threshold is set to detect 4/5th of all forgeries.

Other limitations of the signature verification system include scenarios where some people may have difficulty signing consistently, which can lead to issues with verification. In particular, the pen-up trajectory was found to be a less repeatable signature feature, making it harder to verify some signatures. Although, deleting pen-up trajectories from test and training sets didn't lead to any major improvement, leading the authors to believe that in some scenarios pen-up trajectories are useful.

Secondly, while the model of a person's sign can fit in eighty bytes and can be modified with each use of the credit card, it may not be as accurate as more complex models or algorithms. This could potentially lead to false positives or false negatives in signature verification. Finally, while the algorithm is Immune to forgeries for people who sign often, it may be less effective for people who have more variable signatures or forgeries that are specifically designed to fool the algorithm. Additionally, the algorithm does not depend on the general direction of signing and is insensitive to changes in slope and size, but it may not be able to handle other types of variations in signature style or form.

Offline Writer Recognition Using K Adjacent Segments Vectors [3] presents a method for offline writer recognition through the use of K adjacent segment features in a bag of features framework. The paper aims to tackle the challenge of identifying the author of a handwritten sample when their identity is unknown, by comparing it to a collection of handwriting samples whose authors are already known. The novel KAS feature, models character contours in handwriting. Improving upon previous approaches and achieving a top 1 recognition rate of around 90 plus on the English handwriting dataset (IAM) [6].

The primary limitations of this approach include dataset bias due to the usage of two separate benchmark datasets, limited generalisability of the KAS feature, lack of adaptability to various handwriting styles, lack of focus on forgery detection, and high computational complexity.

3 Methodology

3.1 Patch Scanning

It involves dividing the cheque image into smaller patches and analysing the handwriting in each patch. The goal of patch scanning [1] is to extract features that are essential for the analysis of handwriting and to improve the accuracy of our model. This also enables the model to develop an understanding of the feature representations to expect from various segments of a handwriting sample – from the start of a stroke to the end of one.

Initially, we divide the cheque image into patches of size N x N pixels. We experiment with different values of N to find the optimal patch size that provides the best balance between capturing handwriting details and avoiding noise [9].

In summary, the patch scanning [1] methodology involves dividing the cheque image into smaller patches, which introduces variability in the handwriting samples that undergo analysis. This approach increases the accuracy and robustness of our model and enables us to analyse the handwriting in bank cheques with high precision.

3.2 Image Augmentation

Data augmentation [5] is a critical component of deep learning-based systems, and it has a major impact on increasing the accuracy of such systems. The PyTorch deep learning framework [12] is used to create the image augmentation pipeline. In the case of handwriting analysis, data augmentation [5] can be used to artificially expand the volume of the training set, thereby improving the generalisation capability of the model. In the following section, we delineate multiple data augmentation techniques [5] used in our study.

Rotation

We apply random rotations to the bank cheque images to simulate different orientations of the handwriting. This technique helps the model become invariant to the orientation of the handwriting in the image.

Scaling

We scale the images randomly to simulate variations in the size of the handwriting. This technique helps the model become more vigorous to alterations in the size of the handwriting.

Translation

We apply random translations to the images to simulate alterations in the spot of the handwriting in the image. This technique helps the model become invariant to the position of the handwriting in the image.

Flipping

We flip the images horizontally and vertically to simulate left-handed and right-handed handwriting. This technique helps the model become more robust to variations in the handwriting style.

Distortion

We apply random distortions to the images to simulate the effect of different writing instruments or paper textures. This technique helps the model become more robust to variations in handwriting appearance.

Noise

We add random noise to the images to simulate the effect of different lighting conditions or scanning artefacts [9]. This technique helps the model become more robust to variations in the image quality.

By applying these data augmentation [5] techniques, we are able to increase the volume of our training set, while also improving the robustness of the network [9] to various handwriting styles, sizes, and orientations.

3.3 Feature Extraction Using EfficientNet

The pre-trained EfficientNet [10] deep convolutional neural network is our chosen model for feature extraction from images of bank cheques. EfficientNet is a family of convolutional neural networks that have been optimised for both accuracy and efficiency. We chose EfficientNet because it achieves novel performance on a variety of vision tasks while having a smaller number of parameters compared to other popular networks such as ResNet and Inception.

To extract features from bank cheque images, we fine-tune a pre-trained EfficientNet on a large dataset of handwritten digits and characters [10]. After applying the aforementioned pre-processing, we then randomly divide the dataset into training, validation, and test sets with a ratio of 70:15:15 intro training, validation, and testing dataset.

A cross-entropy loss function is used with stochastic gradient descent optimised using momentum for training the model. The network is trained for 100 epochs and the accuracy of the validation set is measured and used for further fine-tuning (Fig. 2).

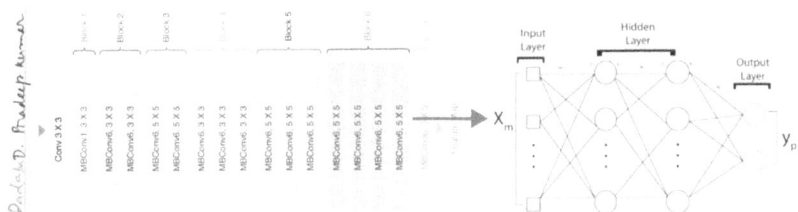

Fig. 2. Model Architecture

After fine-tuning the EfficientNet, we extracted features from each image by removing the final classification layer and using the output of the penultimate layer as the feature vector. This resulted in a compact representation of each cheque image that captured its salient characteristics. These feature vectors were then used for further analysis, such as clustering and classification.

In summary, we demonstrate that fine-tuning EfficientNet for feature extraction [10] from bank cheque images leads to a compact and discriminative representation that can be used for handwriting similarity classification.

3.4 Comparison Using Artificial Neural Networks (ANN)

After extracting the feature vectors of bank cheques using the pre-trained EfficientNet, we need to compare them to verify if they belong to the writer. To accomplish this, we use an artificial neural network (ANN) [8] that takes the feature vectors generated by EfficientNet [10] as input to generate a similarity score between two handwriting patches in the same cheque.

The architecture of the ANN [8] consists of two dense layers with 4096 and 256 neurons respectively. The ReLU activation function is used in these layers. We also add dropout layers after each fully connected layer to prevent overfitting. Finally, we add a sigmoid layer at the end to output a similarity score between 0 and 1, with 0 indicating dissimilar images, and 1 indicating similar images.

During training, we use a binary cross-entropy as the loss function, and the network is trained using stochastic gradient descent optimised using momentum. We train the network on a dataset of labelled cheque images, where each pair of images has a label of either 0 or 1, indicating whether they belong to the same account holder or not. We randomly divide the dataset as training, validation, and test set with a ratio of 70:15:15 into training, validation, and testing datasets.

After training the ANN [8], we evaluate its performance on the testing dataset by computing various metrics, like accuracy, precision, recall, and F1-score. We also generate a confusion matrix [11] to visualise the efficiency of the network on different classes.

In summary, we use an ANN to compare the feature vectors generated by EfficientNet [10] and to determine the similarity between two cheque images. The ANN [8] is trained on labelled images and evaluated using various metrics to determine its accuracy and performance. This approach provides a reliable and efficient way to verify the authenticity of bank cheques and detect any fraudulent activity.

4 Experiments

4.1 Dataset

The primary dataset used for the problem domain is the cheque image dataset called IDRBT [4]. The dataset contains 6,174 images of bank cheques written by 255 different writers from four different banks in India with diverse textures and ink colours. To imitate the pen ink differences within cheque leaves, seven black and blue pens each were used by nine different volunteers to prepare the dataset. Every cheque is written by the volunteers using two different pens, resulting in a dataset with multiple possible combinations of ink. Every check was scanned utilizing a regular scanner at a resolution of around 300 DPI. Additionally, a spreadsheet was included, containing the metadata related to the process of writing and modifying the checks by pairs of volunteers, the pens used by each pair, as well as the specific words written by each volunteer on the checks.

The second dataset we use is the IAM Handwriting dataset [6], which is a benchmark dataset for handwriting recognition and writer identification research. The dataset contains 1,539 handwriting samples from 657 writers, including both English and non-English handwriting. In our model, we only use the English handwriting samples from the IAM dataset [6] for the writer identification task (Fig. 3).

By using two distinct datasets, we aim to train the handwriting recognition model on the IAM Handwriting dataset to increase the model generalisation to variations of a

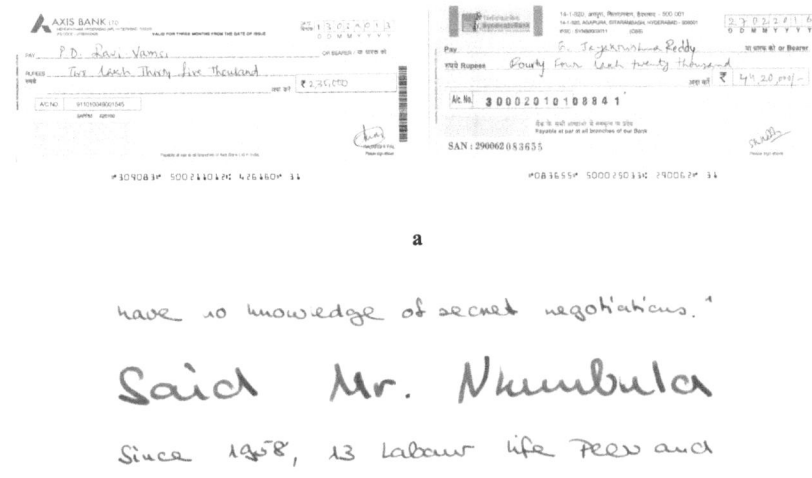

Fig. 3. a. Two input samples consisting of different bank cheque details **b.** Three samples of the IAM dataset

person's handwriting under a wide variety of conditions and then fine-tune the model on the IDRBT dataset suited to our bank cheque verification problem domain. The use of the IDRBT dataset provides a real-world scenario for bank cheque verification, while the IAM dataset [6] provides a platform for building the performance of our proposed technique against existing state-of-the-art techniques in the field of writer identification.

4.2 Experimental Results

Our proposed method gives results that achieve high classification scores of 99.4% and 98.7% accuracy respectively on both the IDRBT and IAM datasets. These scores indicate that the final ANN is able to classify and recognise the relevant features for identifying similarities between handwriting patches.

We assessed the performance of our proposed model on the test set with the following metrics: precision, accuracy, F1 score, and recall. As portrayed in Table 1.

Table 1. Evaluation Metric Comparison

Metric	Value
Accuracy	0.994
Precision	0.984
Recall	0.970
F1 score	0.977

Our model obtained an accuracy score of 0.994, which is higher than the accuracy attained by the latest methods using XGBoost. The precision and recall values were also high, indicating that our model was able to accurately recognise most of the fraudulent cheques while minimising the false positive rate. The F1 score was high, showing a good balance between recall and precision.

To further assess the robustness of our model, we performed a study by deleting each data augmentation technique one at a time and retraining the model. The results are shown in Table 2.

Table 2. Comparison of Accuracy on Evaluation Metrics

Data Augmentation Technique	Accuracy
Baseline (no augmentation)	0.940
Image Rotation	0.965
Image Scaling	0.950
Image Translation	0.955
Image Flipping	0.965
Image Distortion	0.970
Image Noise	0.960

Also, we compared our proposed technique with the baseline method of using PCA for feature reduction and K-means clustering. According to the findings, our proposed approach demonstrated superior clustering performance compared to the baseline method.

Overall, The experimental outcomes exhibit the efficacy of our proposed approach for analyzing handwriting in bank checks by means of clustering. The method can accurately group images based on salient handwriting features, which can be useful in identifying fraudulent cheques.

5 Conclusion

The experimental results showed that the proposed approach of using patch scanning, data augmentation, and feature extraction using EfficientNet achieved high accuracy in handwriting analysis for bank cheque verification. The model performed better with an accuracy of 99% with the English dataset (IAM) [6] and IDRBT data set [4], indicating that it can accurately distinguish between writers for bank cheque verification.

In the paper, we have introduced a novel approach for handwriting analysis in bank cheques using a combination of patch scanning, data augmentation [5], and feature extraction with EfficientNet [10]. Our results demonstrate the effectiveness of our approach, achieving high accuracy and F1 scores for handwriting similarity classification tasks. This work has significant practical implications for the banking industry, enabling automated cheque processing systems to improve their efficiency and reduce errors.

Future work could involve expanding this approach to other handwritten documents, such as forms and contracts, and exploring its potential for signature verification tasks.

Overall, this research contributes to the field of forensic document analysis and could potentially aid forensic document examiners in identifying bank cheque forgeries. Future work can explore the use of other clustering algorithms and deep learning models to further improve the accuracy and efficiency of the proposed approach.

References

1. Xing, L., Qiao, Y.: DeepWriter: a multi-stream deep CNN for text-independent writer identification. In: 15th International Conference on Frontiers in Handwriting Recognition (ICFHR) (2016)
2. Bromley, J., Guyon, I., LeCun, Y., Sickinger, E., Shah, R.: Signature verification using a "siamese" time delay neural network. In: NeurIPS (1993)
3. Jain, R., Doermann, D.: Offline writer identification using k-adjacent segments. In: Document Analysis and Recognition (ICDAR) (2011)
4. Dansena, P., Bag, S., Pal, R.: Differentiating pen inks in hand-written bank cheques using multi-layer perceptron. In: Proceedings of 7th International Conference on Pattern Recognition and Machine Intelligence, Kolkata, India, December 2017
5. Mikołajczyk, A., Grochowski, M.: Data augmentation for improving deep learning in an image classification problem. In: International Interdisciplinary Ph.D Workshop (IIPhDW), Świnoujście, Poland (2018)
6. Marti, U., Bunke, H.: The IAM-database: an English sentence database for off-line handwriting recognition. Int. J. Doc. Anal. Recognit. **5**, 39–46 (2002)
7. Yang, W., Jin, L., Liu, M.: DeepWriterID: an end-to-end online text-independent writer identification system. arXiv (2015)
8. McCulloch, W., Pitts, W.: A logical calculus of the ideas immanent in nervous activity. Bull. Math. Biophys. (1943)
9. Moreno-Barea, F.J., Strazzera, F., Jerez, J.M., Urda, D., Franco, L.: Forward noise adjustment scheme for data augmentation. In: IEEE Symposium Series on Computational Intelligence (SSCI) (2018)
10. Tan, M., Le, Q.V.: EfficientNet: rethinking model scaling for convolutional neural networks. arXiv (2019)
11. Luque, A., Carrasco, A., Martín, A., De Las Heras, A.: The impact of class imbalance in classification performance metrics based on the binary confusion matrix. Pattern Recognit. (2019)
12. Paszke, A., et al.: PyTorch: an imperative style, high-performance deep learning library. In: NeurIPS 2019 (2019)

An Intensive Approach to Solve Linguistic Issue Using Data Mining and Ontology Based Advanced Algorithms

M. M. Uma Maheswari[✉] and N. Arivazhagan

SRM Institute of Science and Technology, SRM Nagar, Kattankulathur 603203, India
{umamahem1,arivazhn}@srmist.edu.in

Abstract. Data mining is a method that is used to uncover previously unknown relationships or patterns in vast amounts of data. This can be accomplished by analyzing the data in question. It is usual practice to gain knowledge from enormous data sets by applying statistical methods. This is done because the data often include intriguing structures within them. In this paper, we develop an intensive approach to solve linguistic issue using data mining and ontology based advanced algorithms. This involves Pre-processing, Feature Selection, Rule Generation, Ontology based classification and Semantic Characterisation. The simulation is conducted to test the efficacy of the model in terms of accuracy and other metrics. The results show that the proposed method has higher rate of accuracy than the other existing algorithms.

Keywords: Intensive Approach · Linguistic · Data Mining · Ontology

1 Introduction

In order to make data mining more efficient, we implement strategies from computer science, machine learning, database technology, and other forms of data analysis [1]. In addition to identifying previously unanticipated structures within the data, it is common practice to derive knowledge from massive data sets by applying statistical methods. In order to improve the efficiency of data mining, we implement many methodologies from the fields of computer science, machine learning, database technology, and other types of data analysis. On the other hand, these typical approaches to data mining pay no attention to the semantics of the data that are being mined [2].

Even though there is a vast selection of statistical functions that may be used to develop models for data analysis, the conceptual and semantic interpretation must still be done by a human who is an expert in the relevant field. This person will unavoidably be required to commit a considerable amount of time to the process of accumulating a comprehensive body of subject knowledge [3]. Researchers working in the field of data mining are confronted with a challenge in this regard since they are required to design ways for incorporating domain knowledge into the process of data mining. Because of the difficulty of this task, there has been an increase in research aimed at inventing efficient ways of data mining, and this pattern is anticipated to continue in the foreseeable future [4].

Researchers have been using the technology of knowledge bases to data mining and machine learning techniques as a response to the considerable increase in the utilization of a knowledge base in applications. This is being done as a direct result of the aforementioned increase. The aim of these initiatives is to solve a wide variety of problems and deficiencies that are inherent in the data mining approaches that are now in use [5].

The conventional approaches to data mining have a propensity to focus on statistical computations while neglecting to take into account the underlying relationships that exist between data pieces. Therefore, the most efficient approach to carry out the necessary data pretreatment tasks, which include filtering, cleansing, converting, and limiting the search space, is not to use systems that are based on statistics [6].

Access to a knowledge source that is unique to the location could perhaps make these actions more successful. A computer system is able to recognize and appropriately manage irrelevant data, impute missing values, and narrow the search space by using the relationship information that is held in the domain knowledge base. This allows the system to make better use of its resources [7].

In the past, it was impossible to acquire these qualities. Second, data mining operations like data normalization, missing value manipulation, feature selection, and selecting a suitable data mining technique for a particular data set require domain knowledge, which is difficult for a nonexpert to acquire. This makes it difficult for a nonexpert to perform data mining operations. It is challenging for someone who is not an expert to carry out data mining operations [8].

The information that is stored in a knowledge base, the process of data mining may be made more efficient and automated, resulting in better overall results. In contrast to tasks that are only concerned with prediction, descriptive tasks have the ability to generate additional insights that can be put into action. Third, the process of analyzing the results of data mining is frequently challenging and time consuming, which makes these results inaccessible to individuals who do not have specific expertise in the field. When a user makes use of a knowledge base, it is feasible to streamline the interpretation of rules and get rid of those rules that are unnecessary to the needs of the user [9].

Because of this, researchers came up with the idea of knowledge-based data mining, which is also known as semantic data mining or semantic machine learning because it is based on ontologies. These sentences make a passing reference to a possible method for data mining in which the ontology and semantics of a certain domain are utilized as a context for the analysis of data pertaining to that domain. This methodology might be used to extract information from large amounts of data [10–13].

2 Related Works

The vocabulary of the traditional thesaurus and the clearly delineated semantic structure of the ontology are astonishingly comparable to one another, the building of the ontology may be carried out with a reasonable amount of simplicity. Several researchers in the academic world have investigated the procedure that must be followed in order to transform the specialized field of the thesaurus into an ontology. Their investigation has mostly concentrated on this procedure. The hierarchy that may be found in the Agricultural Science Thesaurus corresponds to inheritance, property, equivalence, individuals, and various other sorts of relationships [14, 15]. The Food and Agriculture Organization directly translates the hierarchical relationship of the thesaurus to the inheritance relationship of the ontology when they are in the process of converting Agrovoc.

The Unified Medical Language System, sometimes known as UMLS [16], is the end product of combining one hundred different biological vocabularies and classification tables. It is composed of 750,000 concepts, and there are 10 million connections between those concepts. In spite of this, the semantic quality of the UMLS is extremely lacking. At Princeton University, psychologists, linguists, and computer engineers utilized ideas from cognitive psychology to construct Wordnet, a dictionary database that is supported by several languages [17]. HowNet is a commonsense knowledge base that reveals the links between concepts and between concept attributes. It does this by perceiving concepts not as the concepts themselves but rather as description objects. Concepts are represented by Chinese and English words [18].

To realize the objective of constructing an ontology-based electronic government known as OntoGov, the European Union came up with and implemented the E-Europe 2002 action plan [19], which was the first step in the process. Research on the development of an ontology library and a synonym approach for the aim of e-government information sharing has been done by academics hailing from Taiwan who are currently working in China [20].

The standard method of ontology, which is based on the thesaurus, has two problems, both of which are linked to the structure and features of the thesaurus. One of these flaws is that the thesaurus is not always accurate [21]. For one, it is inefficient to rely on the knowledge of domain experts and human resources to manually extract ideas, characteristics, and individuals from a thesaurus in order to establish axioms and inference rules, which in turn restricts the size of the final ontology model. This is because the manual extraction of ideas, characteristics, and individuals from a thesaurus is a time-consuming process that requires human resources. This is due to the fact that manually extracting concepts, attributes, and people from a thesaurus is a time-consuming procedure. Second, it is usual practice to convert thesaurus-based domain ontologies to OWL standards. This is due to the fact that these ontologies already have a predetermined semantic relation [22].

When applied to activities such as semantic search and semantic relation inference, however, this particular form of semantic relation disclosure is fairly fundamental, and it does not possess a particularly potent level of effectiveness. Because the thesaurus is so simple and there are no relationships between the terms, an ontology conversion that is based entirely on a specific domain thesaurus can only provide a rough ontology framework. This does not provide enough information for large-scale semantic search and inference.

3 Proposed Method

This paper makes a contribution to the development of a framework that uses techniques from ML as a mechanism for automatically generating rules. In terms of our methodology, the items listed below are incorporated, as demonstrated in Fig. 1.

An Intensive Approach to Solve Linguistic Issue Using Data Mining and Ontology

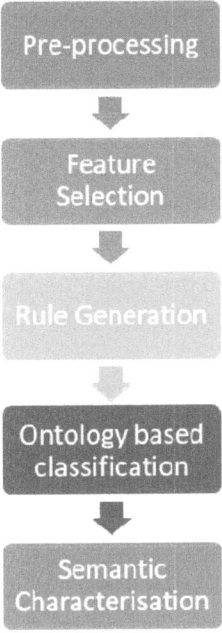

Fig. 1. Proposed Method

Data Pre-processing: During this stage of the process, we bring together the data collected by a variety of sensors, segment the images, and pull out the features. When data from a variety of multi-sensor sources, such satellite photos and LiDAR scans, are merged, a new set of characteristics is formed. These characteristics can be used to better understand the environment. Image segmentation refers to the process of separating individual aspects of an image from the rest of its composition in order to examine it more closely. Each visual object underlying features are analyzed, and the relevance of those attributes is evaluated and scored. Depending on the type of data that is used for fusion, several feature variables, such as spectral and spatial, are obtained. Examples of these variables are spectral and spatial. Spectral and spatial variables are two examples of these types of variables. The extracted feature variables from the multi-sensor data will be output as the result of this module processing. After that, the subsequent module will use these variables as inputs in the calculation that it does.

Feature Selection Based on ML: Since data fusion makes a huge number of features accessible, machine learning (ML) is required in order to choose those features. In this section, we apply machine learning to the process of identifying useful qualities, which is one of the steps in the process. In order to achieve this, we make use of the Boruta method, which is effectively a wrapper around the Random Forest classifier. This allows us to accomplish what we set out to do. Because of this, we are able to determine which variables are of the utmost significance. Utilizing the findings from earlier experiments will allow for the purpose of this work to be accomplished, which is to provide an illustration of the significance of feature selection in multi-sensor data.

Rules Generation Using ML and Ontology: The development of computerized guidelines by the application of machine learning and ontology Our approach to automatically obtaining categorization rules from datasets is predicated on the framework known as inTrees (interpretable Trees), which serves as the basis for our method. Feature selection is where we start with this strategy. The addition of these rules, in addition to the rules that will be produced by specialists in the succeeding unit, will cause an ontology to grow and become more comprehensive.

Ontology Based Image Classification: During the process of categorizing different kinds of photographs, the proposed ontological framework is used. In the course of the investigations, spectral analysis, LiDAR data, and variograms are utilized as sources of information for various qualities.

Semantic Characterisation: The final stage of the procedure is referred to as semantic characterisation, and it entails quantifying the semantic similarities that exist among the several domain classes. A detailed explanation of each of these categories can be found in an ontology. In order to provide a description of domain classes, a semantic variogram is built by employing semantic distances as the building blocks. The semantic variogram is a method that may be used to define class variability by utilizing semantic distances as a measuring tool. This can be accomplished through the utilization of the semantic variogram.

Preprocessing

Dimensionality reduction is the name of one more step that takes place during the preprocessing phase and involves the utilization of ontologies to improve the efficiency of clustering. There are a number distinct approaches one might take in order to compile a fresh data set for use with clustering algorithms. One of these methods involves locating an existing data set that corresponds to the ontology concepts and then extracting relevant information from that set. The fundamental goal of the WordNet ontology is to reduce the dimensions by extracting the nouns from the texts so that they can be utilized as an input data set. This is accomplished by harvesting the nouns from the papers. In addition to that, the ontology was utilized to solve problems concerning polysemy and synonymy by exchanging pertinent ideas for the nouns that were in question. This was accomplished through the use of the ontology. The subsequent phase consisted of applying an information gain assessment to the concepts that had previously been retrieved in order to determine which of those concepts held the most important semantic notions. Following that, the last set of data that was used as input into the clustering method was comprised of these ideas. We utilized Eq. (10), which demonstrates that the information gain is equal to the difference in entropy between a word and the entropy of the extracted semantic ideas, to calculate the information gain. This allowed us to determine how much information we had gained.

$$Gain(n) = e_n(n) - e_c(n)$$

where

$e_n(n)$ - entropy of a noun n, and

$e_c(n)$ - average entropy of the related concepts for noun n.

The entropy of a noun is computed with the following equation:

$$e_n(n) = -\sum_{i \in \pi_n} p(i|n) \log p(i|n)$$

Where

i - cluster,
n - noun,
π_n - noun clusters set, and
$p(i, n)$ - Documents with n of i.

Calculating the entropy of a noun is possible by employing the formula:

$$e_c(n) = -\sum_{c_i \in C(n)} p(c_i|n) \sum_{j \in \pi_s} p(j|n, c_i) \log p(j|n, c_i)$$

where

π_s - semantic cluster set,
$C(n) = \{c_1, c_2, ..., c_k\}$ - associated concept set of noun n, and
$p(i, n)$ - noun n and concept c_i present in a document.

In addition, the symbol $p(i, n)$ denotes the percentage of documents that not only contain the noun n, but also have some sort of connection to the idea c_i. The findings of the empirical research showed that the effectiveness of clustering was greatly improved after the implementation of the technique that had been suggested. We were able to achieve this goal by replacing the terms that were used in the document with the ideas that corresponded to them in MCR 3.0, which is an ontology that was constructed on WordNet. Because of the technique that was put out, we were able to successfully cut down on the number of dimensions that the data set had. We were able to compute the label-document weight matrix by making use of the concepts that were extracted from the data. This matrix may be defined as follows:

$$w_i(t_j, L_h) = \frac{1}{n_t H_D}$$

where

D – Document,
$w_i(t_j, L_h)$ - weight of the label (L_h) for term (t_j),
n_t - cardinality of the labels set for t, and
H_D - cardinality of the terms set in D.

A hierarchical clustering technique was used to determine the topic of the articles by making use of the weight matrices of the terms that were used. The results of this analysis were then presented.

Resolving Linguistic Issues

Semantics are not taken into consideration at all in statistical approaches; rather, these methods are completely reliant on the statistics that are produced from the corpora that are used as a foundation. The bulk of statistical methods, once the language preprocessing stage is complete, largely rely on probability for the early stages of the ontology learning process.

The corpus has been tagged with part-of-speech tags. These tags make it feasible to extract terms and concepts based on the grammatical structures in which they are located. The utilization of this data makes it possible to derive syntactic structures from sentences, such as noun terms and verb terms, for use in further analysis. Discovering terms involves making use of these frameworks and going through the process of parsing the words and modifiers that the frameworks themselves include. For instance, when it comes to ontology learning, the syntactic structure of noun terms, also known as NPs, can be mined for candidate ideas by looking at the NPs themselves. We were able to determine and extract the hypernymous complex terms by employing a syntactic analysis in conjunction with the head-modifier concept. This allowed us to locate the complex terms in which the complex word head operates as a hypernym.

Another term from linguistic theory that can be used in educational activities connected to ontology is the subcategorization frame, and it can be found in the previous sentence. In order to gain insight into the subcategorization frame of a certain word, one can examine the frequency with which that word chooses other words of a particular form.

The action verb to write selects the nouns Bob and letter as its neighbors, which results in the production of a subcategorization framework comprised of the aforementioned three terms. Because of this, from this point forward, the verb write will only be able to select its adjacent nouns from the classes of person and written communication. This limiting of the search, when paired with other methods of grouping, helps find previously obscured aims.

Another typical approach, the use of seed words, is implemented into a variety of the activities that are planned with the intention of teaching ontology. The foundation for other algorithms that extract related terms and concepts is provided by words that are peculiar to a topic and are also referred to as seed words. By using this strategy, the extracted terms are narrowed down to only include those that are conceptually connected to the seed words in the most direct and direct way possible.

Contrastive Feature Selection Analysis

The process of extracting words from a corpus could end up producing outcomes that have nothing to do with the application that was supposed to be done using the corpus in the first place. It is strongly suggested that the advice included in these terms be ignored for the time being. In order to execute the process of deleting extraneous terms from the outcomes of a method for the extraction of terms, a technique known as contrastive analysis is applied. This allows for the task to be completed successfully. Two of the new contrastive analysis metrics that have been established for the subject of ontology learning are known as domain relevance and domain consensus. While one corpus is utilized for the intended domain, also referred to as the target corpus, the other is put to use for all other purposes. After filters are applied to the search results, the terms that are most relevant to the subject domain will be the only ones that are retained.

Examining how relevant a term is to a particular field is one way to determine the level of specialization it possesses in that field. Words receive scores that are determined by how significant they are to the domain that is being targeted in comparison to how unimportant they are to the domains that are being contrasted.

In order for us to achieve this goal, we have collected a list of separate domains, which we shall refer to as domains. For the purpose of determining the relevance $(D_1, .., D_m)$ of the domain in the target domain, D_k, the following formula is utilized:

$$DR = \frac{P(t|D_k)}{\sum_{i=1}^{m} P(t|D_i)}$$

where

$P(t|D_k)$ and $P(t|D_i)$ are used to represent the probability of finding the term t in the domain Dk that is being targeted, and the chance of finding the term t in the domain Di that is being contrasted, respectively.

The formula, which can be found here, can be used to frequency data in order to generate an approximation of this probability.

$$Est(P_t(d)) = \frac{f(t,k)}{\sum_{t \in D_k} f(t',k)}$$

Domain consensus is an alternative that may be used to determine which concepts are shared by numerous publications inside a certain domain, which in this case is Dk. Domain consensus can be used to discover which concepts are shared by several articles. The formula that should be utilized is called.

$$DC = \sum_{t \in D_k} P(t) \log(p_t(d))^2$$

where $P_t(d)$ - probability of t in d of D_k.

A linear combination formula is employed in order to combine the two scores. The notation for this formula is as follows:

$$\text{Score} = \alpha DR + (1 - \alpha) DC$$

where

α - experimental parameter.

Only terms that have an overall score that is higher than a criterion that has been specified are kept.

Classification

We used annotations taken from the Ontology in order to search for genes that carried out functions in a comparable manner. The Ontology is a lexicon that has been standardized for use when talking about genes and the products that genes create.

The organization of ontology is achieved through the utilization of directed acyclic graphs (DAGs), with terms serving as the nodes of the graph and interactions between terms serving as the edges of the graph. Because of the simplicity with which it may be applied and the extensive quantity of biological data that it includes, ontology has developed into an instrument that is essential for the research of gene characteristics.

A semantic similarity measure and an approach known as maximum (Max) combining were utilized in the construction of a functional similarity matrix. Following that, this matrix was utilized in order to compare genes that have been implicated in disease with genes that have not been implicated in disease.

Let imagine that the letter p represents the likelihood of finding the term c in the corpus that we are looking at p(c). Applying the following formula to a term will allow you to calculate its IC:

$$IC(c) = -\log(p(c))$$

Methods that rely on ICs determine the semantic distance between two GO terms by looking up their MICA and calculating the IC of that term. This is how the semantic distance between the terms is determined. The concepts are compared in this manner in order to ascertain the level of semantic distance between them. The metric that was devised by Wang makes use of the taxonomic structure of the GO in order to apply weights to the numerous degrees of semantic similarity that exist between the terms that are entered. This similarity can be found between any two terms that are entered. Max was employed, which entails taking the average of the scores of semantic similarity between all of the GO terms that are connected with a certain set of genes. This was done in order to determine the level of semantic similarity between the two sets of terms.

Let imagine that g1 and g2 are two different genes, and that the GO keywords (t11, t12 ...... t1m) and (t21, t22 ...... t2n) that relate to them are listed below. The maximum combining criteria use the formula that is shown in the following paragraph in order to determine which terms on both lists share the greatest degree of semantic similarity with one another:

$$\max(g1, g2) = \max(t_{i1}, t_{j2})$$

The calculation that was utilized to determine the score for semantic similarity did not take into account GO keywords that contained the electronic evidence code (IEA). The functional similarity of genes was able to be determined for the first time thanks to a language that was derived from various biological processes. On the other hand, there are now gaps in the GO data for multiple genes, which leads to incorrect semantic evaluations. Because the dataset contained genes with a true semantic similarity of zero, our team came to the conclusion that it would be preferable not to assign missing scores a value of zero in order to avoid any possible misinterpretation of the data. This decision was made in order to avoid any potential misunderstandings. Therefore, replacing missing scores with 0 can lead to biased predictions, and it also has the potential to lead to biased evaluations of the performance of classifiers. Both of these problems can be avoided by avoiding the practice altogether.

4 Results and Discussions

The amount of a certain domain that is covered is determined by the completeness measure of the ontological model. One could consider this to be an example of a coverage index. The ontology models that have been shown take into account practically all of the concepts that are used in the process of creating a learner, in addition to the subject matter of a conventional university course, evaluation modules, and an online open course. In addition, important axioms are a component of ontology models. These axioms ensure that there are no redundant or superfluous terms and prevent the insertion of new concepts that fit either category.

Essential components of smart web technology include the ability to analyze acquired ontologies for lexical correctness, idea coverage, taxonomic health, and non-taxonomic appropriateness.

If a proper evaluation of the ontology acquisition process is carried out, the overall ontology learning process can be adjusted to accommodate a situation in which the acquired ontologies do not meet the standards that were established by the user. Because ontology learning is composed of numerous interrelated layers, doing an evaluation of the utility of ontology extraction can be a challenging task to undertake.

In light of the fact that evaluating domain ontologies is a challenging task, numerous assessment strategies have been developed over the course of the past few years. Furthermore, research and development in this field are still ongoing at this time. Each of the

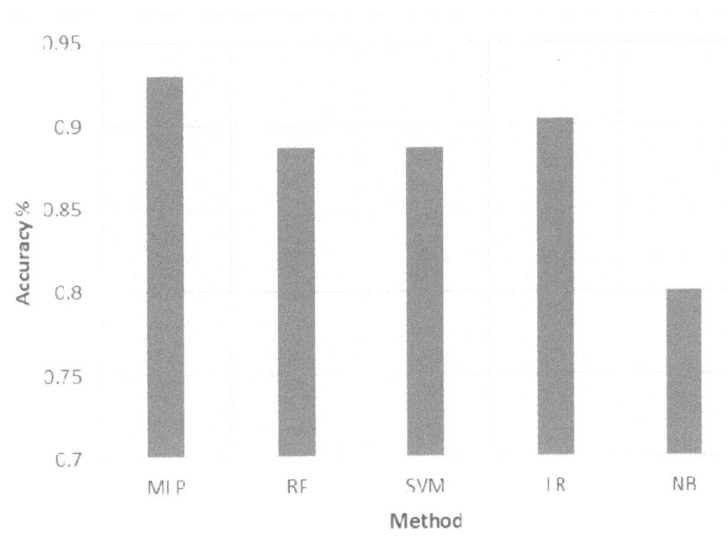

Fig. 2. Proposed MLP method with various ML for Accuracy

methods that have been proposed can be sorted into one of these broader classifications. These classifications are typically arranged in accordance with the various ontologies that the methods were developed to analyze and the goals that they aim to achieve as in Figs. 2, 3, 4, 5, 6 and 7.

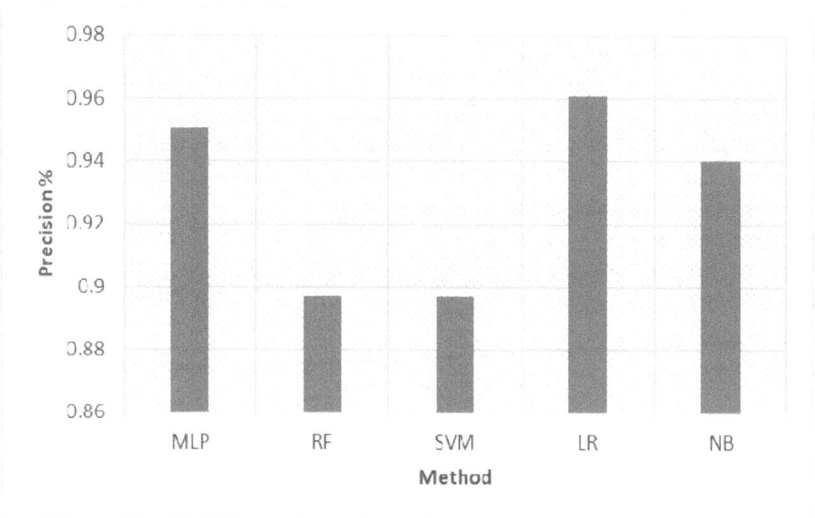

Fig. 3. Proposed MLP method with various ML for Precision

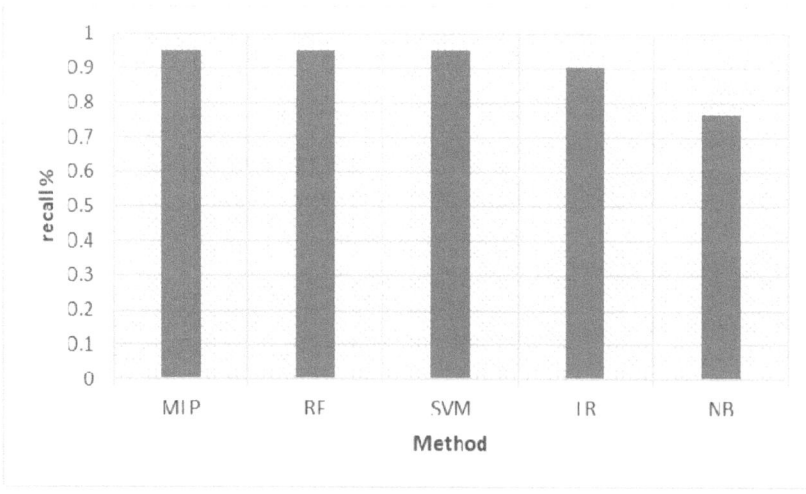

Fig. 4. Proposed MLP method with various ML for Recall

An Intensive Approach to Solve Linguistic Issue Using Data Mining and Ontology   159

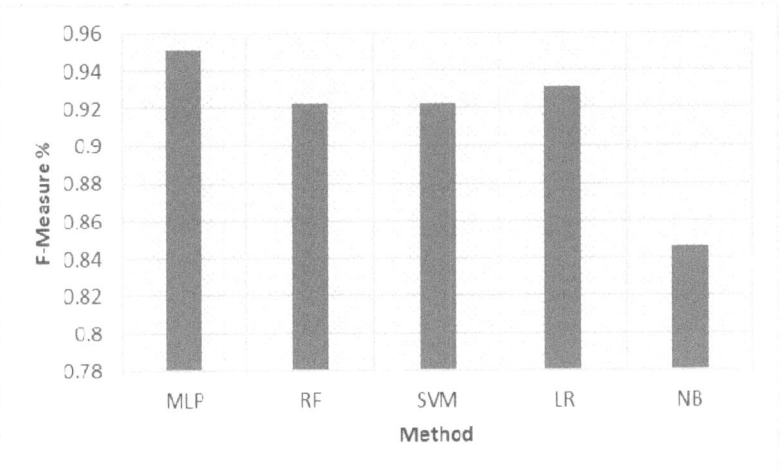

Fig. 5. Proposed MLP method with various ML for F-Measure

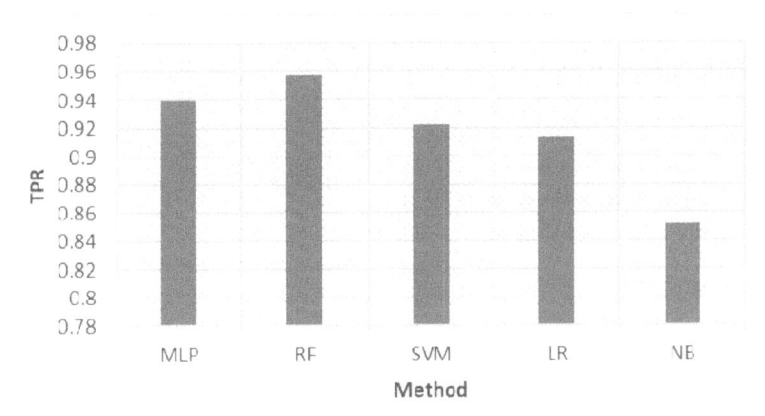

Fig. 6. Proposed MLP method with various ML for TPR

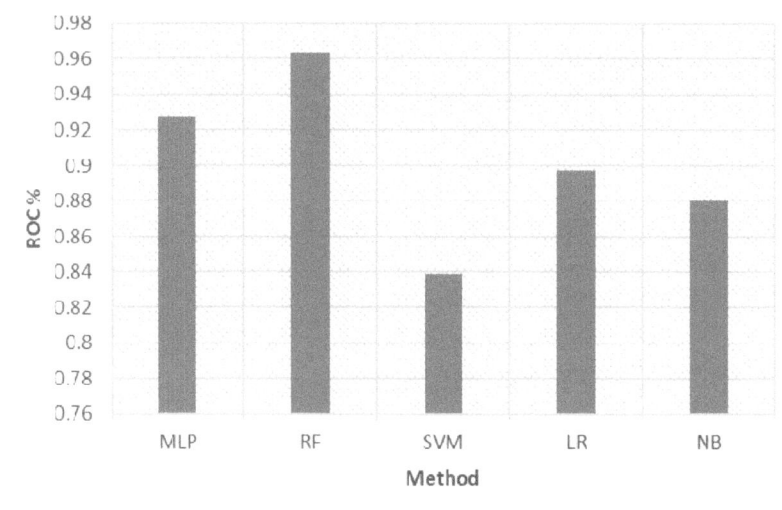

Fig. 7. Proposed MLP method with various ML for ROC

5 Conclusions

In this paper, we create a comprehensive strategy for dealing with linguistic problems by employing data mining and ontology-based sophisticated algorithms. Pre-processing, Feature Selection, Rule Generation, Ontology-based Classification, and Semantic Characterization are all part of this process. Discovering hidden connections or patterns in large datasets is the goal of data mining. It's possible to do this by looking at the relevant data. Making use of statistical methods to glean insights from massive data sets is common practice. This is done because there are often fascinating patterns hidden within the data. The accuracy and other performance metrics are put to the test via simulation. Evidence from testing demonstrates that the proposed approach outperforms state-of-the-art algorithms.

References

1. Hassanien, A.E. (ed.): Machine Learning Paradigms: Theory and Application. Springer, Cham (2019). https://doi.org/10.1007/978-3-030-02357-7
2. Cao, Q., Samet, A., Zanni-Merk, C., de Beuvron, F.D.B., Reich, C.: An ontology-based approach for failure classification in predictive maintenance using fuzzy C-means and SWRL rules. Procedia Comput. Sci. **159**, 630–639 (2019)
3. Meškelė, D., Frasincar, F.: ALDONA: a hybrid solution for sentence-level aspect-based sentiment analysis using a lexicalised domain ontology and a neural attention model. In: Proceedings of the 34th ACM/SIGAPP Symposium on Applied Computing, pp. 2489–2496, April 2019

4. Rajbabu, K., Srinivas, H., Sudha, S.: Industrial information extraction through multi-phase classification using ontology for unstructured documents. Comput. Ind. **100**, 137–147 (2018)
5. Ayadi, A., Samet, A., de Beuvron, F.D.B., Zanni-Merk, C.: Ontology population with deep learning-based NLP: a case study on the biomolecular network ontology. Procedia Comput. Sci. **159**, 572–581 (2019)
6. Salguero, A.G., Medina, J., Delatorre, P., Espinilla, M.: Methodology for improving classification accuracy using ontologies: application in the recognition of activities of daily living. J. Ambient. Intell. Humaniz. Comput. **10**(6), 2125–2142 (2019)
7. Kastrati, Z., Imran, A.S.: Performance analysis of machine learning classifiers on improved concept vector space models. Future Gener. Comput. Syst. **96**, 552–562 (2019)
8. Sulthana, A.R., Ramasamy, S.: Ontology and context based recommendation system using neuro-fuzzy classification. Comput. Electr. Eng. **74**, 498–510 (2019)
9. Liu, J., Li, Y., Tian, X., Sangaiah, A.K., Wang, J.: Towards semantic sensor data: an ontology approach. Sensors **19**(5), 1193 (2019)
10. Hertling, S., Portisch, J., Paulheim, H.: Supervised ontology and instance matching with MELT. arXiv preprint arXiv:2009.11102 (2020)
11. Ali, F., El-Sappagh, S., Kwak, D.: Fuzzy ontology and LSTM-based text mining: a transportation network monitoring system for assisting travel. Sensors **19**(2), 234 (2019)
12. Ozaki, A.: Learning description logic ontologies: five approaches. Where do they stand? KI-Künstliche Intelligenz **34**(3), 317–327 (2020)
13. Wan, C., Freitas, A.A.: An empirical evaluation of hierarchical feature selection methods for classification in bioinformatics datasets with gene ontology-based features. Artif. Intell. Rev. **50**(2), 201–240 (2018)
14. Al-Aswadi, F.N., Chan, H.Y., Gan, K.H.: Automatic ontology construction from text: a review from shallow to deep learning trend. Artif. Intell. Rev. **53**(6), 3901–3928 (2020)
15. Li, G., Wang, Z., Ma, Y.: Combining domain knowledge extraction with graph long short-term memory for learning classification of Chinese legal documents. IEEE Access **7**, 139616–139627 (2019)
16. Liu, B., Yao, L., Ding, Z., Xu, J., Wu, J.: Combining ontology and reinforcement learning for zero-shot classification. Knowl. Based Syst. **144**, 42–50 (2018)
17. Sacha, D., Kraus, M., Keim, D.A., Chen, M.: VIS4ML: an ontology for visual analytics assisted machine learning. IEEE Trans. Vis. Comput. Graph. **25**(1), 385–395 (2018)
18. Natarajan, Y., Kannan, S., Mohanty, S.N.: Survey of various statistical numerical and machine learning ontological models on infectious disease ontology. Data Anal. Bioinform. Mach. Learn. Perspect., 431–442 (2021)
19. Qazi, A., Goudar, R.H.: An ontology-based term weighting technique for web document categorization. Procedia Comput. Sci. **133**, 75–81 (2018)
20. Yuvaraj, N., Kousik, N.V., Jayasri, S., Daniel, A., Rajakumar, P.: A survey on various load balancing algorithm to improve the task scheduling in cloud computing environment. J. Adv. Res. Dyn. Control Syst. **11**(08), 2397–2406 (2019)
21. Konys, A.: Knowledge repository of ontology learning tools from text. Procedia Comput. Sci. **159**, 1614–1628 (2019)
22. Costa, L.A., Pereira Sanches, L.M., Rocha Amorim, R.J., Nascimento Salvador, L.D., Santos Souza, M.V.D.: Monitoring academic performance based on learning analytics and ontology: a systematic review. Inform. Educ. **19**(3), 361–397 (2020)

Author Index

A
Abraham, Pranav 138
Aljanabi, Mohammad 3
Appavu, Narenthira Kumar 24
Arivazhagan, N. 148
Arvind, K. S. 112

B
Babu, Nelson Kennedy 24
Bhandari, Sangya 112

C
Chandra, Pavan 34

D
Dayoub, Ahmad 112

G
Ghandhupu, Poojitha 112
Gopalakrishnan, S. 3, 71
Gupta, Subodhini 131

H
Hasan, Ali 112
Hemanand, D. 3
Hemanth, S. V. 3
Hephzipah, J. Jasmine 3

J
James, Alvin 138
Jeya Aravinth, S. 103
Jeyaram, G. 14

K
Kanagamalliga, S. 43
Karthik, K. 34
Kavitha, S. 43
Kishore, 34

L
Lokesh, S. 55

M
Madheswaran, M. 14
Maheswari, M. M. Uma 148
Malar, R. Shirley Jeeva 14
Manisha, S. 71
Mathavan, B. Anandha 82

N
Neha, 34

P
Paramasivan, B. 82
Priya, K. Sakthi 43
Priya, R. Karpaga 43

R
Rajalingam, S. 43
Rajaprakash, S. 34
Ramesh, S. 138
Ranpariya, Jaydeep 138
Reddy, P. Sreenath 71

S
Salman, Rahama 131
Sarangakrishna, 34
Saravanan, Roshan 138

Shenbagharaman, A. 82
Shunmugapriya, B. 82
Soundari, A. Gnana 3
Sundara Mahalingam, P. 103
Swetha, K. Rani 71

V

Vasantha, S. 55
Vidhya, V. 14
Vishnu, A. 103
Vivekrabinson, K. 103

SPRINGER NATURE

GPSR Compliance

The European Union's (EU) General Product Safety Regulation (GPSR) is a set of rules that requires consumer products to be safe and our obligations to ensure this.

If you have any concerns about our products, you can contact us on ProductSafety@springernature.com

In case Publisher is established outside the EU, the EU authorized representative is:

Springer Nature Customer Service Center GmbH
Europaplatz 3
69115 Heidelberg, Germany

The manufacturer's authorised representative in the EU is Springer Nature Customer Service Centre GmbH, Europaplatz 3, 69115 Heidelberg, Germany. If you have any concerns regarding our products, please contact ProductSafety@springernature.com

Printed and bound by CPI Group (UK) Ltd, Croydon, CR0 4YY

26/03/2026

02078935-0019